Watering White Elephants?

Watering White Elephants?
Lessons from Donor Funded Planning and Implementation of Rural Water Supplies in Tanzania

Ole Therkildsen

Published by
Scandinavian Institute of African Studies, Uppsala 1988

Publications from the Centre for Development Research, Copenhagen

No. 1. Bukh, Jette, *The Village Woman in Ghana*. 118 pp. Uppsala: Scandinavian Institute of African Studies 1979.
No. 2. Boesen, Jannik & Mohele, A.T., *The "Success Story" of Peasant Tobacco Production in Tanzania*. 169 pp. Uppsala: Scandinavian institute of African Studies 1979.
No. 3. Kongstad, Per & Mönsted, Mette, *Family, Labour and Trade in Western Kenya*. 186 pp. Uppsala: Scandinavian Institute of African Studies 1980.
No. 4. Carlsen, John, Economic and Social Transformation in Rural Kenya. 230 pp. Uppsala: Scandinavian Institute of African Studies 1980.
No. 5. Bager, Torben, *Marketing Cooperatives and Peasants in Kenya*. 116 pp. Uppsala: Scandinavian Institute of African Studies 1980.
No. 6. Raikes, Philip L., *Livestock Development and Policy in East Africa*. 254 pp. Uppsala: Scandinavian Institute of African Studies 1981.
No. 7. Therkildsen, Ole, *Watering White Elephants?* 224 pp. Uppsala: Scandinavian Institute of African Studies. 1988.

This series contains books written by researchers at the Centre for Development Research, Copenhagen. Its is published by the Scandinavian Institute of African Studies, Uppsala, in co-operation with the Centre for Development Research with support from the Danish International Development Agency (Danida).

Cover: Painting by Douglas Mpoto, Kinondoni Young Artists, Dar es Salaam.

© Ole Therkildsen
ISBN 91-7106-268-8
ISSN 0348-5676

Printed in Sweden by
Motala Grafiska, Motala 1988

Preface

From 1980 to 1983 I lived in Iringa, doing socio-economic research on water, health and participation. This research was done together with colleagues from BRALUP (later IRA) and CDR. It was part of the preparation of the DANIDA-funded regional water master plans for Iringa, Mbeya and Ruvuma regions. I also visited other regions where donors were supporting the planning and implementation of rural water supplies.

Two problems were evident. The donor funded plans were not much used and the water supply improvements tended not to be sustainable despite relatively large donor inputs of money, material and manpower. It was – and is – customary to blame this on the rapidly deteriorating economic conditions in Tanzania and on inappropriate domestic policies. These were – and are – certainly valid reasons. But I also became increasingly convinced that the donor approaches to planning and implementation contributed significantly to the problems of the non-use of plans and the non-sustainability of schemes. In fact, the donor support to water supply improvements in Tanzania appeared to illustrate the crises of donor funded rural development activities in many African countries very well.

This provided the motivation for the study reported here. It has been financed by a grant from the Danish Research Council for Development Research for which I am grateful.

I would also like to express thanks to many people whose ideas, comments, criticisms and assistance have been especially valuable. Jannik Boesen, CDR and Mark Mujwahuzi, IRA, have both been involved in my work in various ways since 1980. I. Anderson, M. Athanary, W. Balaile, K. Homanen, H. van Schaik and K. Winkel have commented on various versions of the manuscript. B. Lerche, M. Henk, H. Mortensen and K. Wendelboe have managed to make unreadable manuscripts readable. And last, but not least, I wish to thank the very many people in the villages and government and donor offices who have helped in ways too numerous to mention.

Ole Therkildsen

Contents

Preface 5
1. Summary 13
   1.1. Problems 14
   1.2. Findings 16
   1.3. Alternative Approach 18
   1.4. Report Outline 21

Part one: Context
2. The Rural Water Sector: Context and Problems 25
   2.1. Water Supply Problems 26
   2.2. Planning Problems 30
   2.3. Institutional Problems 35
   2.4. Policy Problems 44
   2.5. The International Drinking Water Supply and Sanitation Decade 48
   2.6. The Research Questions and Methods 51

3. Planning and Implementation Approaches 54
   3.1. Approaches and Aid Policies 54
   3.2. Functional versus Normative Planning 56
   3.3. Blueprint versus Process Planning 57
   3.4. Rational-Comprehensive versus Disjointed-Incremental Planning 59
   3.5. Participatory versus Non-Participatory Planning and Implementation 61
   3.6. Bypassing versus Institutional Development in Planning and Implementation 63
   3.7. Two Alternative Planning and Implementation Approaches 67

Part Two: Case Studies
4. The Turnkey "Approach": The Finns in Mtwara-Lindi Regions 71
   4.1. Description of the Project, 1972–1984 71

4.2. Planning 74
4.3. Institutional Framework 75
4.4. Scheme Technology 76
4.5. Service Level 77
4.6. Priorities 77
4.7. Construction 78
4.8. Operation and Maintenance 79
4.9. Participation 81
4.10. Coordination 81
4.11. Monitoring and Evaluation 82
4.12. Key Observations 82

5. The Turnkey "Approach" with a Salesman's Touch:
   The Dutch in Morogoro Region 85
   5.1. Description of the Project, 1978–1984 86
   5.2. Planning 89
   5.3. Institutional Framework 93
   5.4. Scheme Technology 95
   5.5. Service Level 96
   5.6. Priorities 96
   5.7. Construction 96
   5.8. Operation and Maintenance 97
   5.9. Participation 99
   5.10. Coordination 100
   5.11. Monitoring and Evaluation 100
   5.12. Key Observations 101

6. The Pre-Preplanning "Approach":
   The Swedes in Lake Regions 103
   6.1. Description of the Project, 1975–1985 104
   6.2. Planning 106
   6.3. Institutional Framework 108
   6.4. Scheme Technology 109
   6.5. Service Level 110
   6.6. Priorities 111
   6.7. Construction 111
   6.8. Operation and Maintenance 112
   6.9. Participation 113
   6.10. Coordination 114
   6.11. Monotoring and Evaluation 115
   6.12. Key Observations 116

7. The Project Cycle "Approach":
   The World Bank in Mwanza Region   118
   7.1. Description of the Project, 1977–1984   118
   7.2. Planning   120
   7.3. Institutional Framework   120
   7.4. Scheme Technology   121
   7.5. Service Level   121
   7.6. Priorities   121
   7.7. Construction   122
   7.8. Operation and Maintenance   123
   7.9. Participation   123
   7.10. Coordination   125
   7.11. Monitoring and Evaluation   125
   7.12. Key Observations   125

8. The Village-by-Village "Approach":
   The Danes in Iringa, Mbeya and Ruvuma Regions   127
   8.1. Description of the Project, 1979–1985   128
   8.2. Planning   130
   8.3. Institutional Framework   135
   8.4. Scheme Technology   140
   8.5. Service Level   140
   8.6. Priorities   141
   8.7. Construction   141
   8.8. Operation and Maintenance   142
   8.9. Participation   144
   8.10. Coordination   145
   8.11. Monitoring and Evaluation   146
   8.12. Key Observations   147

Part Three: Interpretations
   9. Achievements and Failures of Control-Oriented
      Planning and Implementation   153
      9.1. Classification of Case Studies   153
      9.2. Achievements and Failures   155
      9.3. Persistence of Control Orientation   161

   10. Limits of Control-Oriented Planning and Implementation   162
      10.1. Political and Bureaucratic Conflicts   162
         10.1.1. The Recipients   162
         10.1.2. The Donors   165

  10.1.3. The Beneficiaries 167
  10.1.4. Goal Displacement 167
 10.2. Multiple Decision Makers 168
 10.3. Uncertainty and Complexity 170
  10.3.1. Unpredictability 170
  10.3.2. Poor Knowledge 170
  10.3.3. Inadequate Information 172
  10.3.4. Planning without Implementation 174
 10.4. Aid Without Village Level Roots 175
  10.4.1. Lack of Mobilization 176
  10.4.2. Lack of Empowerization 179
  10.4.3. Constraints on Participation 180
 10.5. Aid Without Institutional Roots 181
 10.6. The Limits: A General Trend 185

Part Four: Alternatives
11. Rural Water Supply Development Through Experiments and Learning 189
 11.1. Development Projects as Policy Experiments 191
 11.2. The Learning Process Approach 196
 11.3. The Adaptive Approach 199

12. The Adaptive Approach to Planning and Implementation of Rural Water Supply Development in Tanzania 201
 12.1. Sector Policies and Donor Assistance 202
 12.2. Introducing Adaptive Planning and Implementation 203
 12.3. Increasing Capacity to Plan and Implement from Village Level and Upwards 206
 12.4. The Dilemma 208

References
 Published 210
 Unpublished 217

List of Abbreviations 223

List of Tables

Table 1 RWMPs: Donors, Consultants, Costs and Implementation 33
Table 2 Resource Allocations to Rural Water Supplies in Tanzania, 1964–1984 46
Table 3 Rural Water Supply Baseline Data. Global Surveys, 1970–1985 49
Table 4 Major Decade Constraints in Developing Countries. Global Survey, 1983 50
Table 5 Major Assumptions and Characteristics of Control Oriented and Adaptive Planning and Implementation Approaches 66
Table 6 Finnish Funding and Output of the Mtwara-Lindi Projects, 1972–1987 72
Table 7 Dutch Funding and Output of the Morogoro Projects, 1978–1984 88
Table 8 Swedish Funding of the Lake Region Projects 105
Table 9 World Bank Funding and Output of the Mwanza Project, 1979–1983 119
Table 10 Danish Funding and Output of the Iringa, Mbeya and Ruvuma Projects, 1979–1988 129
Table 11 Summary Assessment of Control Oriented Planning and Implementation Approaches 158
Table 12 Direct and Total Costs of Wells Constructed in Donor Assisted Projects 183
Table 13 The Experimental Approach 192

List of Figures

Figure 1   The Case Study Regions in Tanzania   14
Figure 2a General Structure of the Ministry at Central Level, 1972–1984   36
Figure 2b Overall Government Organization at District, Regional and National Level, 1972–1984   37
Figure 3a Flow Chart for Budget Preparation of Rural Water Sector before 1984   40
Figure 3b Planning and Monitoring Process Involving Government and Party before 1984   41
Figure 4   The Separation of Functions in Control Oriented Planning and Implementation   174
Figure 5   Users versus Government Locations of Domestic Water Points   178
Figure 6   Schematic Representation of Fit Requirements in the Learning Approach   197

1. Summary

Rural drinking water supply came high on the agenda in Tanzania in 1971 when the Party declared that by 1991 all rural households should have access to safe water within easy reach of their homes. The provision of water schemes and the cost of construction and maintenance were to be a government responsibility. No major official changes in approach have been made since then.

Foreign donor agencies have played significant roles in the planning and implementation of this policy. From 1974 to 1983 Regional Water Master Plans (RWMPs) were prepared in 17 out of Tanzania's 20 regions. They are medium- to long-term sector plans, typically with a time horizon of twenty years, in which the "optimal" use of all water resources in the region is specified, with particular emphasis on "firm guidelines" for water supplies for human and livestock consumption. All but one were funded by aid agencies at a cost of USD 1 to 2 mill. per region. Donors have also assisted in the preparation of similar plans covering only parts of a region or covering only specific activities such as the construction of wells. All these plans were mainly prepared by foreign consultancy companies and academic institutions.

Some donors subsequently got involved in the implementation of water supply projects in twelve regions with typical funding levels of around USD 0.5 to 1 mill. per region per year. This aid was frequently channelled through donor controlled regional implementation units. On average, donors have provided around two-thirds of the total capital expenditure for rural water sector development over the last 15 years.

This study focuses on the preparation of donor funded medium- and long-term rural water supply plans and their subsequent implementation. It is mainly based on case studies of the involvement of five donors: Denmark, Finland, Holland, Sweden and the World Bank. These donors have operated in half of Tanzania's regions (see Figure 1). The case studies cover various periods since the mid-1970s. Field work was completed in 1985. The study does not include the planning and implementation of water supplies in urban areas. Sanitation and health education activities are also excluded.

1.1. Problems

There are two main reasons for undertaking this study. Despite the fairly substantial donor support of funds and technical assistance to the planning and implementation of rural water supplies, it is commonly agreed that the medium- and long-term plans are only used to a

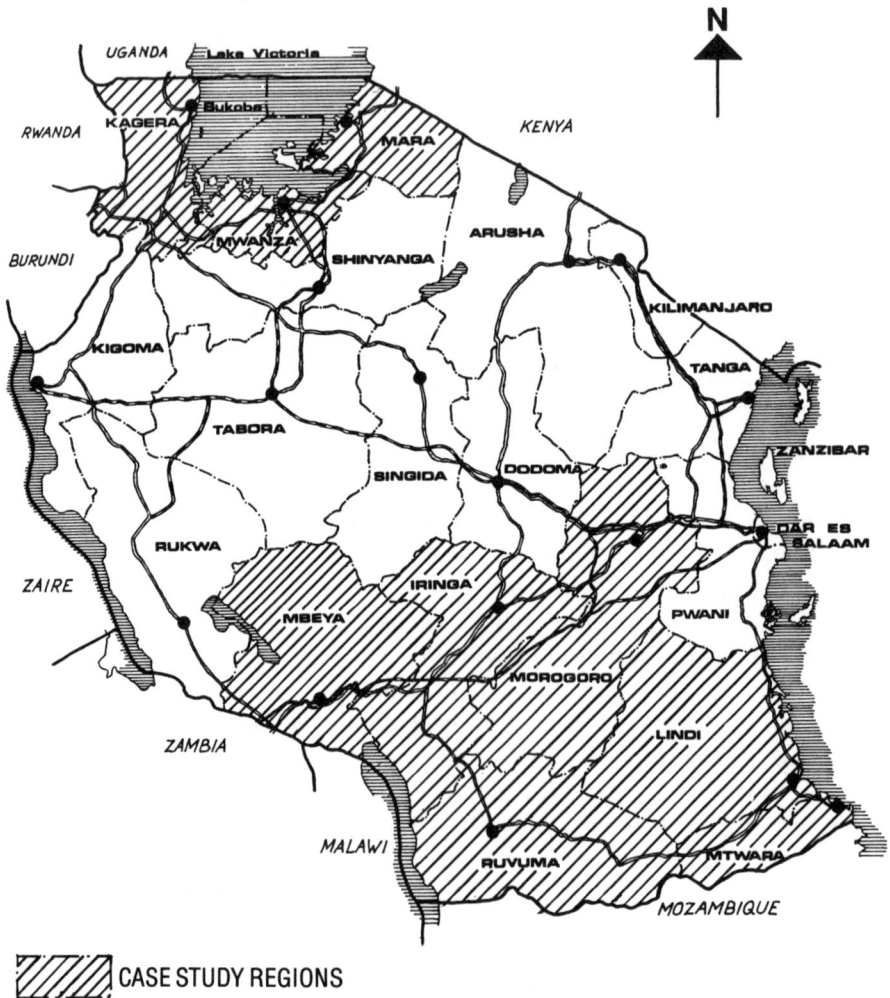

Figure 1: *The Case Study Regions in Tanzania*

limited degree during implementation – or not at all. In particular the non-engineering aspects of the plans have been ignored during implementation (training, participation, institutional development). In the cases where plans have been used, the emphasis has typically been on the purely technical aspects, and on the fast production of new water schemes. Likewise, surveys show that the benefits from completed schemes are not sustained. Schemes often cease to function soon after being handed over to their users and the Tanzanian authorities. Hence the main purpose of the study is to identify those characteristics of the donor planning and implementation approaches which in themselves have contributed to the non-use of plans and the non-sustainability of implemented activities.

The study is mainly written for planners, policy makers, administrators, and researchers dealing with donor involvement in rural water sector activities in particular and with rural development in general. In the Tanzanian context the study will, hopefully, contribute to the ongoing discussion of the role of donors in rural development activities. However, Tanzanian experiences are not unique. Other developing countries have embarked on similar ambitious rural water supply programmes during the International Drinking Water and Sanitation Decade, 1981–90. With the Decade emphasis on water to all by 1990 and the stress on extensive planning and donor support to reach this goal, there are considerable similarities with the approach that Tanzania embarked on already in 1971. Tanzania can therefore in many ways be regarded as a "pilot country" for Decade activities.

Perhaps the experiences from the Tanzanian rural water sector have even wider relevance. The specific conditions under which the planning and implementation of such activities are carried out here, do not differ significantly from the general conditions for rural development in many sub-Saharan countries. This context is characterized by significant complexity and uncertainty with respect to resource availability, settlement patterns, institutional linkages and capacities, peasant responses to government initiated activities, actual and proclaimed rural development policies, and political support for rural development. Therefore this study may be relevant for planning and implementation of a number of rural development activities.

1.2. Findings

The planning and implementation undertaken by the five donors included in this study have been *control-oriented*, albeit to varying degrees. The general features of this approach are (i) a focus in the medium and long-term plans on a set of future construction targets typically geared towards the 1991 goal; (ii) a fairly detailed pre-implementation specification of the means to reach the goals, indicating, for example in five-year phases, village priorities, supply technologies, organizational arrangements, manpower and financial needs, implementation rates, etc.; (iii) substantial collection and analyses of information prior to implementation on the basis of which the plans are specified; (iv) an implicit or explicit specification in the plans of the role of the intended beneficiaries either as passive receivers of services, or as participants in various pre-determined activities, sometimes followed by attempts to mobilize beneficiaries to participate in these activities during implementation; and (v) a bypassing of recipient organizations at national, regional and district levels by the technical assistance teams, especially during the preparation of medium- and long-term plans but also to a certain degree during implementation.

Obviously there are some variations among the specific donor approaches. But in general the case studies illustrate that the medium- and long-term plans have not been used very much during implementation, although the preparation of plans may have induced donors to increase their assistance to the rural water sector. The case studies illustrate the non-sustainability of the Finnish, Dutch and World Bank assistance given prior to 1984. Finland and Holland are now changing their approach. The World Bank has completely withdrawn.

In contrast, the Danish approach is characterized by a certain degree of user participation which follows procedures based on direct implementation experience and specified in the RWMP. The Swedish approach has recently undergone a complete change and is now also directly linked to active user participation. However, in both the Danish and the Swedish cases the degree of participation is still fairly limited, and long-term experiences of its impact do not yet exist. The same holds for the recent Swedish attempt at implementing with a lesser degree of bypassing of local institutions than used by other donors.

A number of factors account for the non-use of plans and the non-sustainability of rural water sector activities. Obviously the ever-deepening crisis in the Tanzanian economy since the mid-1970s has seriously affected all development activities. It has a detrimental impact on government finance, foreign exchange availability, administrative capacity, and so on. However, the control-oriented planning and implementation approach contributes significantly to the problems of donor assisted sector activities, because the approach is based on five key assumptions that do not hold or only hold to a limited extent.

In the uncertain and complex rural water sector context it cannot be assumed that the donor, the recipient and the beneficiaries agree on and are committed to a set of common objectives on which medium- and long-term plans can be based. Differing and conflicting agendas are typical, and conflicts tend to occur during the implementation rather than the planning stage.

The implementation of control-oriented plans also assumes the existence of one or a few authoritative and powerful decision-making agencies with the means to enforce compliance. Such agencies do not exist in the complex institutional setting in the sector, which reflects the general weakness of the Tanzanian state.

Attempts at detailed pre-implementation planning of medium- and long-term activities presuppose the predictability of future economic, institutional and political conditions, etc.; knowledge about the interrelation between future inputs and outcomes; and the existence of operational information about key sector activities. Yet unpredictability, lack of knowledge and operational information are typical of the rural water sector and its environment – especially in the pre-implementation period.

Participation tends to play a minor role in control-oriented planning and implementation. This is partly because user participation may conflict with the desire by donors and the recipient to control activities from above; and partly because it is assumed that planned activities will fit beneficiary needs, and that user acceptance, resource commitments and knowledge can be mobilized at will if and when needed. But the case studies illustrate that planners and implementors can neither exert strong control over villages, nor assure the utilization and functioning of water supplies without active user cooperation.

Finally, bypassing of local institutions is based on the assumption that the lack of recipient capacity to plan and implement can be efficiently substituted by technical assistance staff in the short run without serious long-term consequences. However, the case studies indicate that despite high implementation rates, donor-dominated implementation is rather costly. Worse still, it tends to divert attention and resources from the need to build up domestic institutional capacity towards the production of new schemes. Bypassing, furthermore, tends to fragment and limit (perhaps even destroy) the long-term capability of recipient organisations to take over donor-assisted activities.

Clearly, spending more resources on pre-implementation planning, specifying future activities in more detail, and increasing donor influence in the implementation stage are not solutions to the problems of non-use of plans and non-sustainability of activities. It is the control-oriented approach itself which is at fault – although its negative effects are amplified by the deteriorating economic conditions in Tanzania.

1.3. Alternative Approach

The uncertainty and complexity of rural water sector activities and the context in which they are carried out require a more adaptive approach to planning and implementation. This differs from the control-oriented approach on five key points – several of which have an important time dimension, as will be discussed in the concluding remarks.

Development projects and programmes should be regarded as policy experiments. Rather than the control-oriented emphasis on detailed medium- and long-term plans, the adaptive approach emphasizes the formulation of long-term strategies and policies to guide short-term planning and implementation. However, the strategies are temporary. They change with experiences and power constellations among participating agencies and beneficiaries.

Planning without implementation leads to implementation without planning. These two functions are therefore linked in the adaptive approach. The organizational separation of planners and implementors which is typical in the control-oriented approach is diminished or abolished. Likewise the strong emphasis on very detailed and comprehensive pre-implementation planning is substituted with a continuous planning-implementation-planning process facilitated by the organizational merging of these two functions.

Errors in planning and implementation are important sources of learning. Hence the linkage between these two functions and hence the importance of an information system in the adaptive approach. The main purpose of the information system is not – as in the control-oriented approach – to monitor deviances between what was planned and what was implemented to help assure plan compliance. The purpose is rather to encourage error detection, to learn from errors, and to make adjustments in plans or implementing activities on the basis of this experience. An information system is therefore of key importance in the adoptive approach. The beneficiaries are important sources of and users of information in this system. But apprehension with regard to detection of errors by donor and host management and staff must be overcome to make it work.

This leads to the fourth key feature of the adaptive approach: the strong emphasis on a continuous dialogue with the intended beneficiaries. The control-oriented approach is biased towards service provision, with participation as an ancillary activity. This contrasts with the adaptive approach where active user participation is regarded as a precondition for external assistance, and as an integrated part of its planning and implementation. For not only does participation help to adjust such assistance to user needs and local knowledge. It also makes it possible to match activities to the willingness and capability of users to commit their own resources in cash or kind to them. However, participation of this kind requires that agency procedures and structures are made to fit participatory decentralized planning and implementation.

Finally, the adaptive approach is based on the assumption that the utility of a plan and the sustainability of activities are at least as dependent on the capacity of organizations at various levels to make it work as it is on the plan design itself. The emphasis on plan documents and on visible results of donor-assisted activities that are typical of the control-oriented approach, lead to excessive pressures for fast results – in this case measured in completed water schemes – and makes it difficult to move beyond a welfare approach to rural development assistance. It is much easier, faster and controllable to construct schemes than it is to build up recipient capacity to maintain them. The slow process of capacity building is a key feature of the adaptive approach. It requires a long-term commitment to task-oriented training, linked to implementation experiences at village level. It requires that user capacity to participate in planning, construction, operation, maintenance and monitoring of activities is increased. And at district, regional and national level it necessitates that existing skills are enhanced and that new ones are introduced to fit the adaptive approach to planning and implementation.

Capacity building also requires that the prime role of technical assistance staff is changed from that of a "performer" or "substitute" working outside recipient agencies, to that of a "teacher" or "mobilizer" working within them to enhance their ability and commitment to implement sector strategies. This, in turn, may require that donor assistance is not channelled exclusively into projects, but also into programmes, and that it is adjusted to the capacity of recipient organizations to absorb the external resources provided.

However, adaptive planning and implementation will not alone ensure the long-term sustainability of rural water sector activities. Obviously they will collapse if the sector is starved of resources or if the economy at large deteriorates significantly.

Adaptive planning and implementation does not mean that planning is de-emphasized and replaced with murky generalities. But the approach does imply that planning and implementation activities should basically match the capacity of local organizations to carry them out. Donors should assist in increasing this capacity – not substitute for it. Thus, if organizational capacity is limited, detailed planning and fast implementation rates through technical assistance support should be avoided. But as recipient capacity and experiences grow, so will the possibility for and desirability of more detailed pre-implementation planning and faster implementation rates.

The time dimension is therefore very important in relation to specific planning and implementation requirements. This also indicates that the adaptive approach cannot be introduced overnight. The key consideration is how planning and implementation tasks can be matched with the recipient capacity to carry them out in the various regions and districts; and how donors can match their assistance with the various needs for local capacity building.

In Tanzania it will be impossible to maintain the past years' emphasis on the fast construction of new schemes if donors and recipients change their approach. Here lies the major dilemma. Through the control-oriented approach, donor assistance may reach more people in the short run, but the long-term sustainability of activities is questionable. The adaptive approach may slow down the speed with which the recipient and the donor reach the target groups. The long-term sustainability of such assistance is, however, more likely, because the external assistance is also based on local commitments and capacity to plan and implement.

1.4. Report Outline

The context and the theoretical framework of the study are presented in Part One. Chapter 2 is specifically written for readers who are not familiar with the rural water sector problems or with Tanzania. Chapter 3 contains a brief discussion of five major dimensions in various approaches to rural development planning and implementation.

In Part Two (Chapters 4 to 8) the five case studies are presented – each chapter being structured in the same way. The title of each chapter captures a distinct feature of the planning and implementation approach used by the particular donor: Finland, Holland, Sweden, the World Bank and Denmark. Each case study can be read on its own. The Dutch, the World Bank and the Danish cases represent the main variations in the approaches. The Finnish and the Swedish cases are to some extent derivations of approaches used by the Dutch and the Danes respectively.

The interpretation of the case studies are given in Part Three (Chapters 9 and 10). First the achievements and failures of the donor approaches are assessed. Then – in Chapter 10 – the reasons for the failures of the control oriented approach are linked to the five faulty assumptions on which this approach is based.

An adaptive approach to planning and implementation is presented as an alternative in Part Four, Chapter 11. Two variations of this approach are discussed. In Chapter 12 it is proposed how the recipient and the donors may gradually introduce this adaptive approach in their ongoing rural water sector activities in Tanzania.

Part one
Context

2. The Rural Water Sector

During the 1970s many donor and recipient countries began to channel resources into activities aimed at poverty-alleviation in rural areas. This gave birth to a large number of donor-funded integrated development activities, basic needs programmes and similar rural projects. Typically the stated aim of such activities was to increase agricultural productivity; to expand employment opportunities; to provide greater access to social service activities for the poor – or all of these.

It soon proved difficult to plan and implement such activities. The rural poverty problems are complex and the solutions uncertain. There were many failures. Plans were not carried out as prescribed. The activities implemented were not benefiting the poor as intended. Consequently, several donors and recipient governments began to put more resources into the planning and management of these poverty-oriented rural development activities in order to reduce the risk of failure.

The activities in the rural water sector in Tanzania are a good example of this trend. They started when the improvement of water supplies became a real issue in 1971, following the Party declaration that all rural people should have easy access to safe water before 1991.[1] All costs involved in the construction, operation and maintenance of water schemes were to be a government responsibility.

This policy – together with other rural development policies based on the Arusha Declaration of 1967 – attracted support from numerous donors. Since 1971 donors have been involved not only in the financing, but also in the planning and implementation of rural water supplies, as discussed later.

The task implied by the 1971 policy was then – and remains today – a colossal one. Just consider its physical magnitude. Tanzania covers an area more than twice as large as Sweden. The variation in rainfall and water resources is considerable. The rural population is settled in

---

[1] The Party Declaration has been written into the technical design criteria for water schemes. "Easy access" means that households should have no more than 400 metres to walk to a water post. "Safe water" is based on WHO's water quality criteria, adjusted to Tanzanian conditions.

more than 8000 widely scattered villages and communication between them and the district and regional towns is often slow, unreliable and costly. Around 85 per cent of the total population of 19.5 million people stay in the rural areas. Their number doubles every 20 to 25 years. To keep up with population growth, more than 400,000 additional people must be supplied with safe water each year. If the 1991 goal is to be reached, more than 1.5 mill. people must be supplied each year from 1986 – assuming that 40 per cent of the full capacity has already been installed[2] (Balaile, 1983). The various technical, social, economic, institutional and political problems involved in the planning and implementation of rural water sector activities in this context are discussed below.

2.1. Water Supply Problems

Seven selected water supply problems are specifically dealt with in each of the five case studies presented in Chapters 4 to 8. They illustrate many of the important issues in the planning and implementation of rural water supply improvements.[3] Here the general background to these seven problems is given.

Scheme technology is a critical issue. In Tanzanian villages three types have been commonly used: piped schemes in which diesel pumps are used to lift the ground water; piped schemes where water flows by gravity from rivers and streams to lower lying villages; and hand-pumped schemes where shallow groundwater is the source. Over the years – and particularly in the 1980s – there has been a shift away from diesel pumps which are expensive to construct, operate and maintain. However, the criteria for choosing appropriate scheme technologies are unclear. No common cost calculation methods are adhered to. The financial cost of a handpump well, for example, varies from Tsh 20 to 70 thousand depending on the calculation method used (Hordijk *et al.*, 1982). This makes the concept of "low-cost" less than useful as a criterion for technology choice. Common criteria for calculating economic costs are also lacking. There is a similar lack of a standardized approach to spare parts. And the organizational implications of various technologies are rarely considered. In the absence of clear Tanzanian directives (see Ch. 2.4), the question facing donor planners

---

[2] This assumption is questionable, as discussed in Chapter 2.1.
[3] The list is not complete. Water resource availability, for instance, is not included. This issue falls outside the scope of this study

and implementors is how to balance the obvious need for common guidelines and coordination with the mercantile interest of individual donor countries for particular technology and brand choices.

Public water posts are the basic feature of Tanzanian village water supplies. House connections are not a usual feature. The *service level* is therefore commonly expressed as the number of people sharing a water post, or as the maximum walking distance between a post and the homes of the users. The Tanzanian design standard stipulates a maximum distance of 400 metres. It has remained unchanged since the early 1970s.[4] Taken as a guideline for initial planning of schemes such figures are useful. Taken as rules, they may result in widespread non-use of water posts because various surveys show that (i) women and children – the main drawers of water – seek to minimize water collection efforts.[5] They almost always prefer the nearest source to the house, unless it is badly polluted, water is scarce, or its use restricted; (ii) numerous traditional water sources (ponds, streams, etc.) exist in most villages, and many of them can be more convenient to use than water posts (BRALUP/CDR, 1982, Ch. 8); and (iii) availability of traditional sources often changes drastically with the season so that access to them at times can be easier than to public water posts. Water use patterns are, therefore, complex. Public standposts compete with traditional sources in attracting users. This poses a dilemma for planners and implementors. If the water use patterns of women and children are ignored, water posts will not be used as intended (IDRC, 1985; Hannan-Andersson, 1985). If scheme designs (e.g. location of water posts) are to be based on detailed information on water use patterns, it would require time-consuming, extensive and costly surveys. The case studies illustrate various approaches to this problem.

One important approach deserves special attention. Knowledge about water use patterns can be obtained from the villagers themselves. This raises the issue of user *participation* in rural water supply development. On the one hand participation can be regarded as an essential part of planning and implementation approaches. On the other hand participation can be a goal in itself. Both aspects are implied in Tanzanian policy statements and in numerous donor proclamations (Feachem, 1980 and Ch. 2.4). But even if the declarations on

---

[4] There is one exception. In 1975, as a result of the rapid and countrywide movement of rural people into village settlements, the Party declared that all villages should be provided with one good source of water by 1981. However, this goal has only had a limited impact on actual planning and implementation except in 1975 and 1976.
[5] The standard work on water use studies is White, Bradley and White (1972).

participation are taken at face value, significant planning and implementation problems exist. The challenge is to make participation operational. How are rural water supply activities matched with beneficiaries' needs? How and to what extent can users be involved in providing information, labour, resources, etc? Should special efforts be made to involve women? How can communication between users and government agencies be organized? What are the rights and duties of users vis-à-vis the government bureaucracy? These questions relate to planning and implementation methodology and are discussed further in Ch. 3.5.

Increased user participation is now also regarded as an important factor in solving *operation and maintenance* problems. Miller (1979, 133), for instance, concluded on the basis of a large survey that:

> Self-help and participation had their most powerful impact on the operations and maintenance aspects of water systems. This is most important because this is usually the weakest area of rural water supplies.

However, poor functioning of schemes remain *the* key sector problem in Tanzania. Numerous surveys conducted over the last ten years have indicated that many schemes, even new ones, function imperfectly or not at all (Engström and Wann, 1975; WHO/World Bank, 1977; Mujwahuzi, 1978; TISCO, 1980; CCKK, 1982b, Ch. 6.3). By 1985, completed schemes had the capacity to serve around 40 per cent of the population but due to non-functioning schemes and low utilization only 10 to 20 per cent of the rural population are actually served (Therkildsen, 1986). Increased participation is, however, no panacea. Operation and maintenance problems are very complex and involve both technical, economic, political and institutional issues that require solutions today as urgently as they did fifteen years ago. More of the background to these issues is provided in the sections below.

The problems of keeping completed schemes running are, furthermore, closely related to the scale, pace, organization and funding of *construction*. Should the planned rate of production of new schemes be determined by (i) official Tanzanian policy statements on targets for sector expansion? (ii) availability of domestic and external funds? (iii) domestic administrative capacity to construct new schemes? (iv) domestic administrative capacity to maintain completed schemes? (v) donor willingness and ability to substitute for perceived domestic shortcomings in financial, technical or administrative capabilities? or (vi) willingness and ability of users to contribute to construction or maintenance? The answers will influence the donor's approach to construction with respect to duration of support, implementation rate, organizational arrangements, and the role of the intended beneficiaries. The case studies illustrate the various donor approaches.

Another set of issues relate to the *priority criteria* according to which candidate villages for new schemes are selected. Tanzanian policies are ambiguous on this. Both criteria related to economic development and need have been discussed. It has been suggested that first priority should go to villages with the highest agricultural potential; to villages with the highest incidence of water related diseases; to villages with the longest distance to reliable, safe and adequate water sources; or to villages which could be supplied at the lowest cost (BRALUP/CDR; 1982, Ch. 12). Without specific Tanzanian guidelines, donor planners and implementors have either left the selection of priority villages to ad hoc decisions by the Tanzanian political-administrative system (see Ch. 2.3), or established their own individual criteria, as the case studies will show.

Monitoring and Evaluation systems are normally regarded as key elements in planning and implementation of rural development projects (Gow and Morss, 1985). They are needed to obtain information about sector activities on the basis of which activities can be managed on a continuous basis; plans can be adjusted; and goal-achievements can be measured. In Tanzania it is a perennial problem that very little reliable and relevant information on water sector activities at village level ever reaches the district and regional government agencies or the central authorities. The lack of information about the functioning of water schemes is a good example of this. District and Regional Water Engineers are supposed to report on this on a quarterly basis. But in their reports they simply assume that all schemes work and that each scheme is used by all families in the villages where water schemes have been installed (BRALUP/CDR, 1982, Ch. 6.3.2).

It is an additional problem that reporting within and between government agencies is limited. Required reports are not produced in time. A management consultant found, for instance, that often it is not possible for the Ministry to follow sector development "because some regions fail to submit their quarterly reports in time".[6] But even with the reports prepared in time

---

[6] (AIB, 1979). AIB found that in 1977/78, 7 out of 20 Regional Water Engineers did submit reports to the Ministry in time for them to be used in the annual reports on the sector.

it is doubtful if they could be used for effective performance control . . . The comments to the reported figures are brief and insignificant; the reports cover implementation of new projects only; and cost figures for operation and maintanance are never reported. (AIB, 1979, 32)

Crozier (1967, 185) has defined bureaucracy as "a system of organization which cannot correct its behaviour by learning from its own mistakes. The process of feedback: mistake – information – correction is not functioning." According to this definition the Tanzanian agencies responsible for water activities are indeed bureaucracies. Unfortunately, as the case studies will show, donor planners and implementors have their own bureaucratic problems. There is little evidence that they have introduced viable operational monitoring and evaluation systems in the regions where they have worked.

2.2. Planning Problems

Each case study (Ch. 4 to 8) also deals with the problems related to the planning of the rural water sector activities funded by donors. Here the general background is provided.

From a planning point of view a report by Rimer (1970) is important. This report was commissioned to prepare for continued Swedish rural water sector support in the 1970s, and to advise on the consequences of stepping up sector activities to reach full coverage in the rural areas within 40 years as specified in the Second Five Year Plan (1969–73).

Rimer and his associates did not work in a vacuum while preparing their report. They were strongly influenced by G.F. Lwegarulila who, one year later, was to become the first Principal Secretary in the new Ministry of Water Development and Power. Lwegarulila was not only an outspoken and competent advocate of rural water supply development but also had unusually good contacts within the Party. One of the conclusions reached by Rimer and advocated by Lwegarulila was truly amazing. It would be possible to provide all rural people with access to safe water within 20 years twice as fast and at only 15 per cent of the cost envisaged in the Second Five Year Plan *if* some major changes were made. They included (i) establishing "user associations" in the villages; (ii) introducing a user water rate to pay for sector development; (iii) establishing a new parastatal to be in charge of sector development; (iv) making RWMPs without foreign assistance;

and (v) preparing each RWMP in stages over 20 years so that implementation experience could be used in plan preparation on a continuous basis.

Rimer's report was presented to the government in April 1970 and made public later that year. Helped by strong supporters in the bureaucracy and by the prevailing post-Arusha Declaration enthusiasm for rural development it was well received. In 1971 the government, the Party and Parliament adopted the goal of water for all by 1991. But only two of the major changes proposed by Rimer in order to reach this goal were adopted in a modified form. A new Ministry of Water Development and Power was established in 1971 – not a parastatal organization as recommended by Rimer. Furthermore, the idea of water master-planning caught on although the specific proposals by Rimer did not (see Boesen, 1986).

The major features of the water master-planning approach were shaped during an influential seminar on rural water supplies for East Africa held in Dar es Salaam in April 1971. Here a wide range of local and foreign practitioners and academicians discussed the type of planning needed to reach the 1991 goal. The importance of comprehensiveness was clearly stressed in the summary of the discussions.

> Water is only one of many development activities all of which must be considered together. Agriculture, settlement programmes, cattle, marketing, all influenced the demand for water . . . and are in turn closely affected by water schemes. (Tschannerl, 1971, 4)

There were also some speculations on the manpower requirements for this. Berry and Conyers (1971, 43), for example, stated that:

> . . . the range of skills needed could be met by a team including the following specializations: hydrology; engineering economics; geography or land use studies; and regional planning.

It was likewise argued that foreign assistance was needed to do this planning within a few years. Thus the Minister of Water stated (BRALUP, 1971, 8):

> The Ministry will, in the next 3 to 5 years, complete the preparation of a water master plan for the whole of mainland Tanzania. This will be done largely with the technical assistance of a number of friendly countries.

That donor involvement in the preparation of RWMPs may also increase the amount of aid for the subsequent implementation was a strong motive in government circles as well at that time, although it was not explicitly stated.

Finally, the need for coordination of these foreign teams was stressed:

> It will be important to have a carefully designed general format within which all plans should be drawn up, particularly since they are to be prepared by ... different teams, most of whom are ... relatively unfamiliar with Tanzanian conditions. (Berry and Conyers, 1971, 35)

By November 1971 the Ministry's Annual Technical Conference recommended that Regional Water Master Plans should be prepared for all regions within three years. The Ministry in particular and the other relevant Tanzanian institutions in general were ill-prepared for the massive donor funded planning effort which was set in motion by this decision. The ambitious 1991 target added to the problem. Foreign aid on a massive scale was clearly needed to reach this target. It made the Tanzanian authorities scramble for donors. There were only a few attempts to guide the RWMP exercise and to integrate it in the ongoing activities.

Thus, no attempts were made to match donors to particular regions. The Ministry had no explicit regional priorities. Therefore donors appear to have had a free hand in picking regions. The order in which the RWMPs were made (see Table 1) does not seem to follow any clear pattern.

Donors were also free to pick consultants to prepare the RWMPs – subject to Tanzanian approval. Invariably the bilateral donors selected consultants from their own countries, although this sometimes created conflicts with the Ministry (see Ch. 8). The foreign consultants had very little prior experience with water master planning. Tanzania was quite simply a pioneer in this field. Thus, an interesting case of reverse transfer of knowledge has taken place. With respect to the Nordic countries, Wingaard (1983, 31) has summarized the situation as follows: "The amount of total planning expertise we have with respect to water master-planning is to some extent also gained through our extensive work in developing countries."

All consultants signed contracts with the donor and not with the Tanzanian authorities. At least formally the latter therefore had to deal with the consultants through the respective donors. And when

Table 1. *RWMPs: Donors, Consultants, Costs and Implementation*

RWMP Funder	Regions	RWMPs prepared by	During	Foreign cost[a] (mill.Tsh)	RWMP implementation
Canada	Dar es Salaam, Coast	CBA Engineering Canada	1976–78	37	None
Denmark	Iringa, Mbeya, Ruvuma	CCKK and CDR, Denmark; BRALUP Tanzania	1980–83	41	Technical assistance and funds since 1980
Finland	Mtwara, Lindi	Finnwater Finland	1973–76	37	Technical assistance and funds since 1976
Holland	Shinyanga;	Nedeco, Holland	1971–73	16	Technical assistance and funds 1974–77
	parts of Morogoro	DHV Consulting Holland	1978+	n.a.	Technical assistance and funds since 1978
Japan	Kilimanjaro	Japanese Int. Coop. Agency	1975–77	n.a.	None
Norway	Rukwa, Kigoma	Norconsult and Chr. Michelsens Inst., Norway BRALUP, Tanzania	1980–82	25	Technical assistance and funds since 1980
Sweden	Mwanza, Mara Kagera	Brokonsult Sweden	1975–1977	32	Technical assistance and funds since 1983
Tanzania	Dodoma	Min. of Water	1971–73	n.a.	None
West Germany	Tanga	Agrar and Hydrotechn., W. Germany	1974–76	12	Implementations as part of rural integrated dev.project since 1977
World Bank	Tabora	Brokonsult Sweden	1978–79	18	None

a Direct costs only. Figures based on Balaile (1983), Kleemeier (1982) and own estimates derived from government budget figures. Direct local costs amounting to Tsh 28 mill. cannot be broken down into regions.

disagreements become serious, formal channels *are* used. In this way donors have had considerable influence over the entire RWMP exercise as the case studies will show.

The Terms of Reference (TOR) for the RWMPs were also strongly influenced by the donor agencies. Each agency appears to have written its own TOR – subject to Ministry approval. Therefore the content of the TORs vary significantly from region to region (Helland-Hansen, 1983, 65–73). "It was normally a matter of reading through them and then approve", according to a Tanzanian official close to the process. Like their consultants the donor agencies themselves had either limited or no experience at all of water master-planning at the time they had to write the TORs. Despite this, it has been a general feature of the TORs that they have been specified in detail *prior* to the arrival of the RWMP planners in the field. Few, if any, substantial changes have been made in them during the planning process itself. The TORs have been blueprints for the RWMPs themselves.

All this has made the coordination of the various RWMPs very difficult and only a few attempts at coordination have actually been made – and made too late. Around 1972 a RWMP coordinator was placed in the Ministry and funded by UNDP (the lead agency for the International Drinking Water Supply and Sanitation Decade). This function remained a one-man job until 1980, when a Water Master Planning Coordinating Unit (WMPCU) was established – mainly financed by the Nordic countries. By that time twelve RWMPs had been completed and the planning teams for the next five regions were already in the country. Little was left for them to coordinate. Furthermore, the unit has had difficulties in establishing itself within MAJI. As an ordinary section within the Project Preparation division in the Ministry, it is placed fairly low in the organizational hierarchy (see Figure 2a). Here it has had insufficient status to formulate and enforce guidelines for the RWMP exercise or the subsequent implementation. When it tried to do this, there were immediate conflicts with the other divisions. The unit – mainly staffed by expatriates and fairly inexperienced counterparts – has therefore had a very limited impact (Athanari *et al.*, 1983). By 1985 the last Nordic donor (Norway) decided to withdraw from the WMPCU. The unit is now staffed by Tanzanians, but its influence remains modest.

But perhaps the most obvious problem with the RWMPs from an implementation point of view is the lack of formal procedures and legal provisions for their use. "Approval" of a RWMP by the donor and the Tanzanian authorities merely implies that the consultant who prepares the plan has complied with the terms of reference for the task. It does *not* mean that the two parties agree with the plan proposals made by the consultants. In fact the two parties may still differ on the extent to which they would implement the various RWMP proposals despite their "approval" of the plans as such. The government-to-government agreements or the "agreed minutes" of meetings between the two parties which normally follow when RWMPs have been approved, do not solve the problem. Such agreements are usually formulated in rather general terms. As guidelines for implementation they are not sufficiently specific and operational.

The basic problem with the RWMPs is therefore a general one for long-term planning. This type of planning is simply not institutionalized in Tanzania despite the emphasis on it.[7] This not only means that such plans have an uncertain status. It also means that Tanzania has no established framework and procedures for linking the RWMPs to the other types of ongoing planning as will be discussed below.

2.3. Institutional Problems

Donor-funded planning and implementation takes place within a complex Tanzanian institutional setting. The specifics for each donor project are described in the case studies. Here the general background is presented. The description refers to the period from the early 1970s to 1984 when District Councils were reintroduced.[8]

Figure 2a shows the structure of the Ministry at the central level. The Ministry has four main responsibilities with respect to water activities: (i) to prepare long-term development strategies including the preparation of RWMPs and their coordination through the WMPCU as already discussed; (ii) to conduct sectoral manpower training; (iii) to plan and implement major water-related activities, the so-called "national projects" of which the preparation of RWMPs is one example and the construction of larger water schemes is another;

---

[7] The plethora of long-range plans produced in Tanzania all suffer from this uncertain status. At the national level these include: five-year plans; a 20-year perspective plan; a structural adjustment plan; a new economic survival plan; a sanitation master plan; a transport master plan; and a plan for "Health for all by the year 2000"; etc. At the regional level there are integrated rural development plans and district water master plans, etc. At the multi-regional level they include: the Uhuru Corridor regional physical plan, the Rufiji Basin Development Plan; and the Kagera Basin Development Plan.

[8] The District Councils had not yet taken over water sector activities by the end of 1985. They are therefore not discussed here.

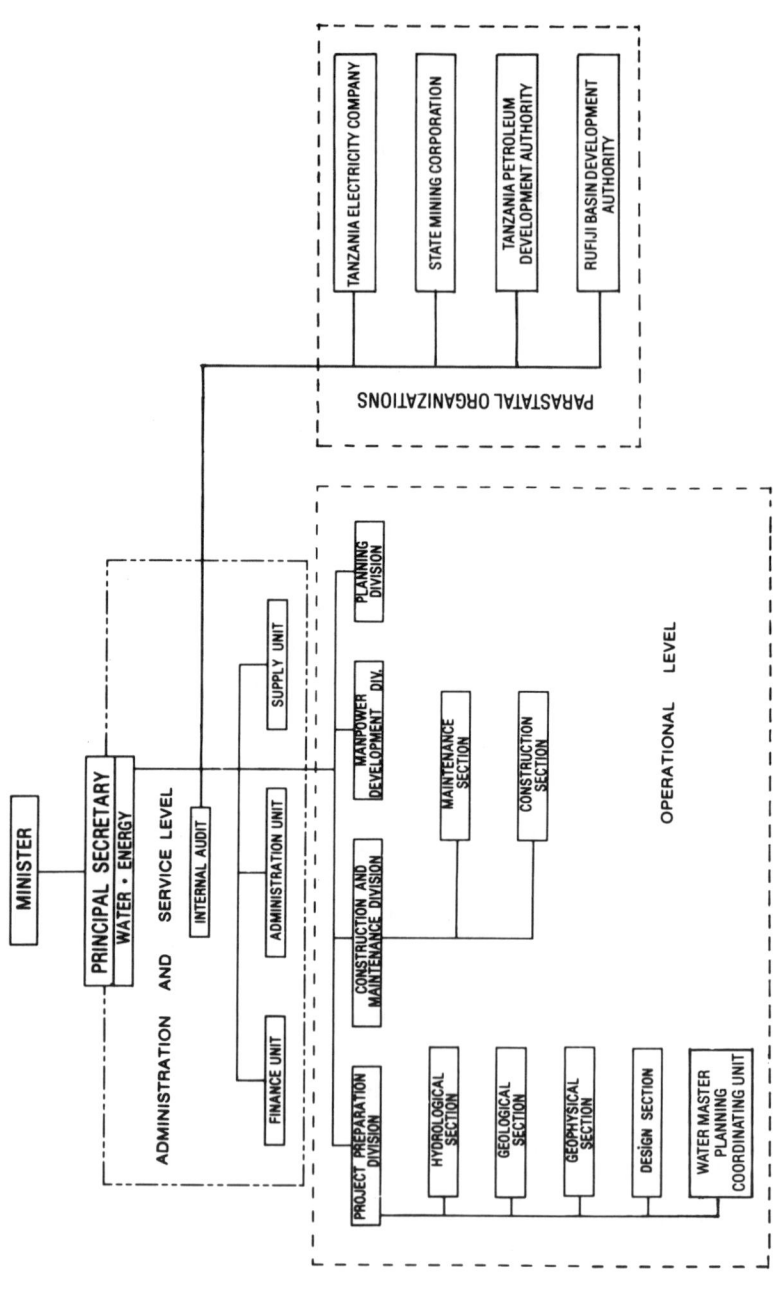

Note: *The Ministry has changed name 4 times since 1971. The figure illustrates key pre-1984 features*
Sources: *WHO/World Bank (1977, Annex 5) and Mascarenhas (1983, 366)*
Figure 2a: *General Structure of the Ministry at Central Level, 1972-1984*

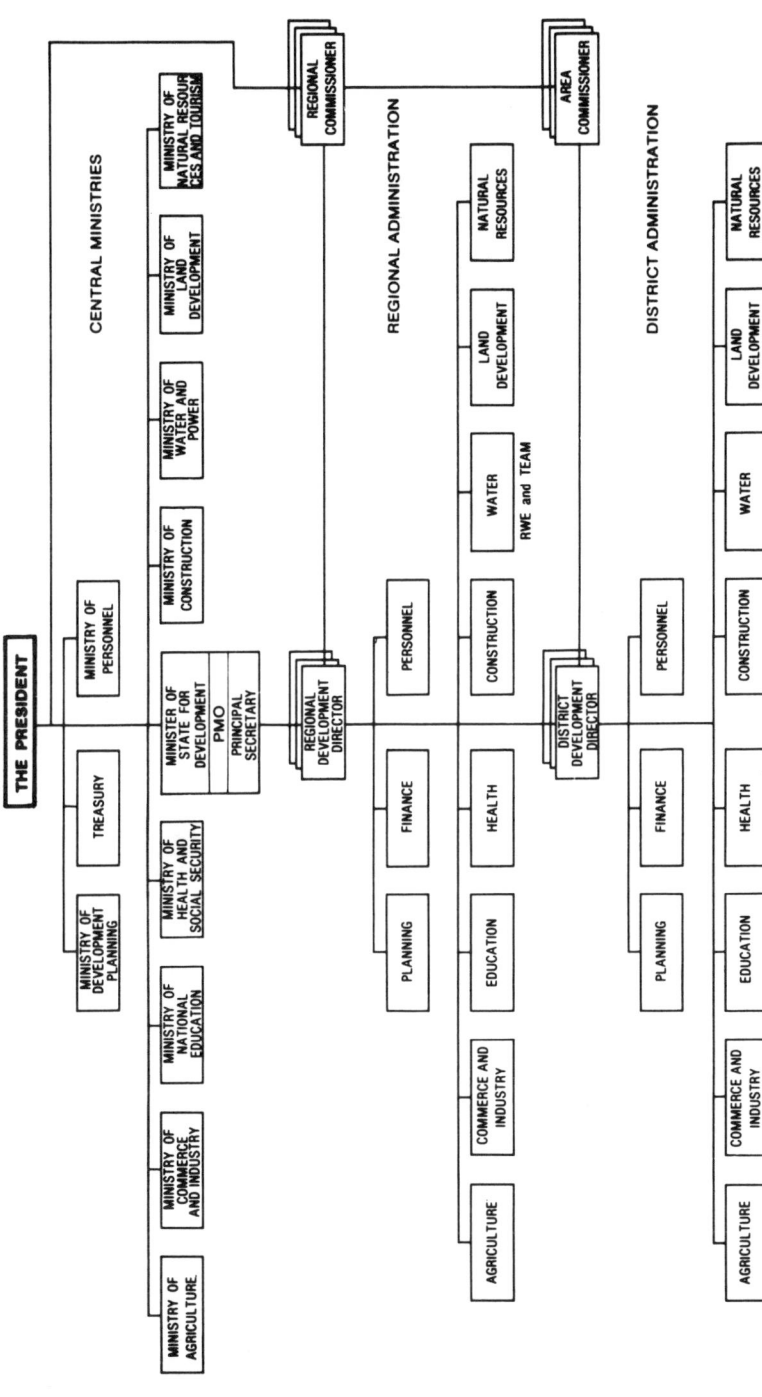

Note: *The Ministry has changed its name 4 times since 1971*
Source: *Mascarenhas (1983, 340)*
Figure 2b: *Overall Government Organization at District, Regional and National Level, 1972-1984*

and (iv) to provide technical advice and backup to the regions on "regional" projects requiring smaller capital investments.[9]

Figure 2b shows the overall organizational structure of the government from 1972 to 1984: the central ministries, and the regional and district administrations. Since the decentralization in 1972 – which was actually a deconcentration – substantial planning and implementation powers have been moved to the regions and districts. A matrix structure with dual reporting and responsibilities was introduced at the same time (Mushi, 1978; Maeda, 1983).

On technical matters concerning "regional" projects, the Ministry is to provide the Rural Water Engineers (RWEs) and District Water Engineers (DWEs) with advice and sometimes approval of planning, construction and maintenance activities. The RWEs and DWEs, in turn, are in charge of the actual implementation of such regional activities. However, the DWEs tend to deal mainly with maintenance activities and minor construction work.

Administratively the RWEs and DWEs report to the Prime Minister's Office (PMO) through the Regional and District Development Directors respectively (RDDs and DDDs). These Directors are personally responsible for achieving the development objectives of their regions and districts. They have the authority to coordinate development planning in their area and to manage the implementation of completed plans. They are accountable to the President through the Principal Secretary, PMO. The RDDs and DDDs delegate their responsibilities for water activities to the RWEs and DWEs respectively.

The procedures for planning and implementation within this set-up are complicated. Figure 3a shows a simplified flow chart for the preparation of the yearly budget. This budget is the only institutionalized and binding planning instrument in Tanzania (Waide, 1974). Its preparation is a mixture of top-down and bottom-up processes, the details of which are explained in Maeda (1983) and Mascarenhas (1983).

Figure 3a shows that the budget guidelines and priorities are issued to the PMO (for regional projects) and to the Ministry (for national projects). Based on these, the two ministries make their annual budgets and plans. However, at the PMO the budget ceilings and priorities are reconciled with the project proposals from the regions. In

---

[9] The definitions of regional and national projects are ambiguous. In 1982 projects with a capital expenditure on less than 1 mill.T.Shs were regarded as "regional". However several small projects may be aggregated to a "national" project.

principle they originate from the villages. The figure also illustrates that the regional projects may not always be forwarded to the Ministry of Water for technical review.

But this is not the complete picture. At every level in the organizational hierarchy the Party is involved in both planning and implementation. Figure 3b shows how proposals *in principle* originate from the village level and then pass through not only technical committees (the management teams to the right in Figure 3b), but also committees with Party representation (the Development and Executive Committees to the left in Figure 3b). The end station for this bottom-up flow is *in principle* the Economic Committee of the Cabinet (a government organ) and the National Executive Committee (a Party organ). During implementation the committees at each level will *in principle* monitor the planned activities.

As the case studies show (and as discussed in Chapter 3.6), many donors have bypassed this political-administrative system both in the planning and implementation phase. They have often established their own project management units at regional level. This is, indeed, a tempting solution since administrative problems abound. Complicated or inappropriate procedures constitute one major problem and their negative effects on performance have been amplified by the growing economic crisis. There are many examples. When funds for planned activities can only be secured for one year at a time and next year's budget may be drastically different from this year's; when the purchasing power of funds has shrunk in real terms (see Table 2); when the availability of materials and equipment has been drastically reduced; and when procedures for procurement, storage and requisition are very cumbersome indeed, it is not surprising that the capacity of the RWEs to construct schemes was reduced by half between 1975 and 1979 (Brokonsult, undated).

Present procedures contribute to other problems as well. The dissimilar procedures of regional and national projects weaken the possibility for central coordination of and support to regional activities. (Regional projects account for around 30 to 50 per cent of yearly capital expenditures for new projects.) More seriously, capital and recurrent budgets are prepared separately and independently of each other. Finally, it is obvious that the long, complicated and byzantine procedures for bottom-up participation in development activities illustrated in figure 3b tend to defeat their own purposes (Finucane, 1974; Mushi, 1978; Holmquist, 1979; Fortmann, 1980; BRALUP/CDR, 1982).

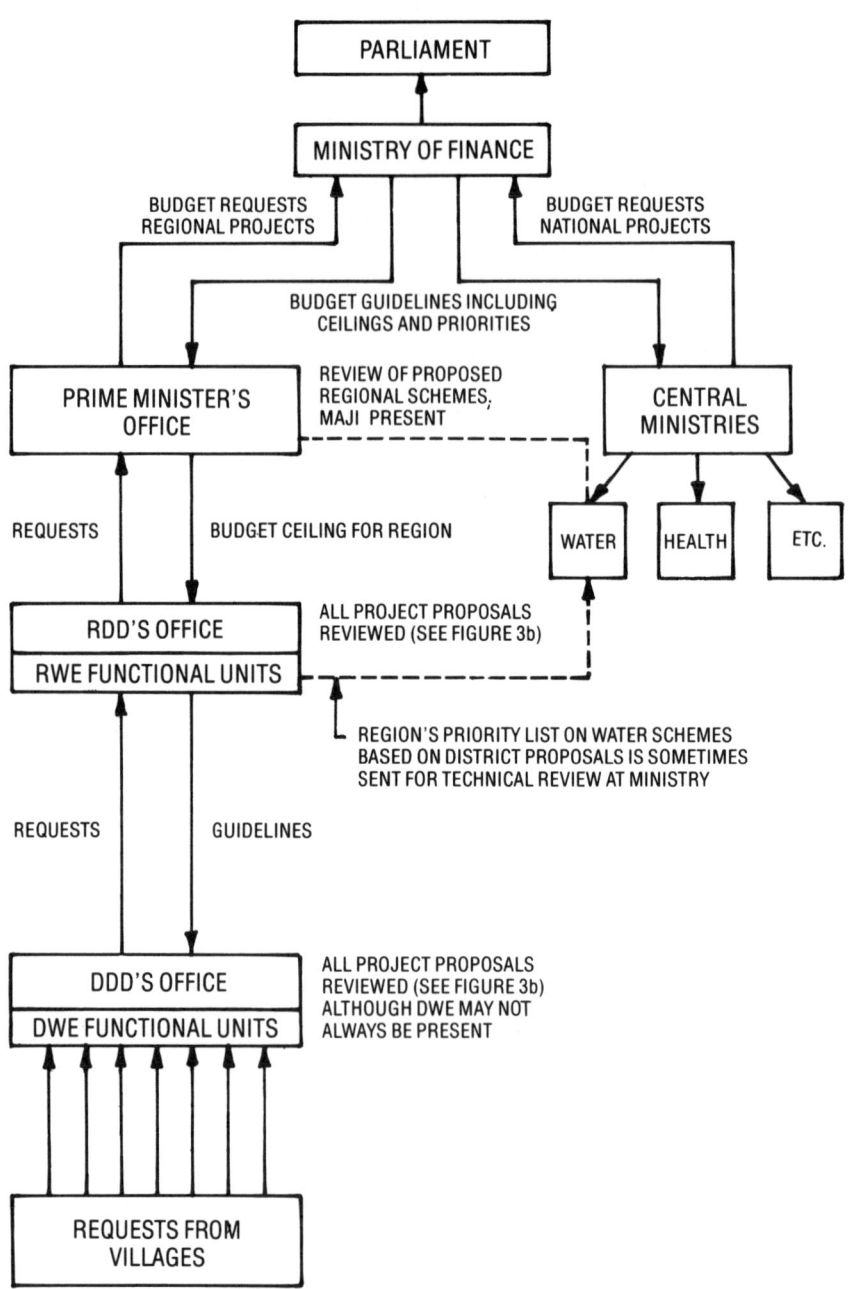

Sources: *WHO/World bank (1977) and Mascarenhas (1983, 342)*
Figure 3a: *Principles for the Budget Preparation of Rural Water Sector before 1984*

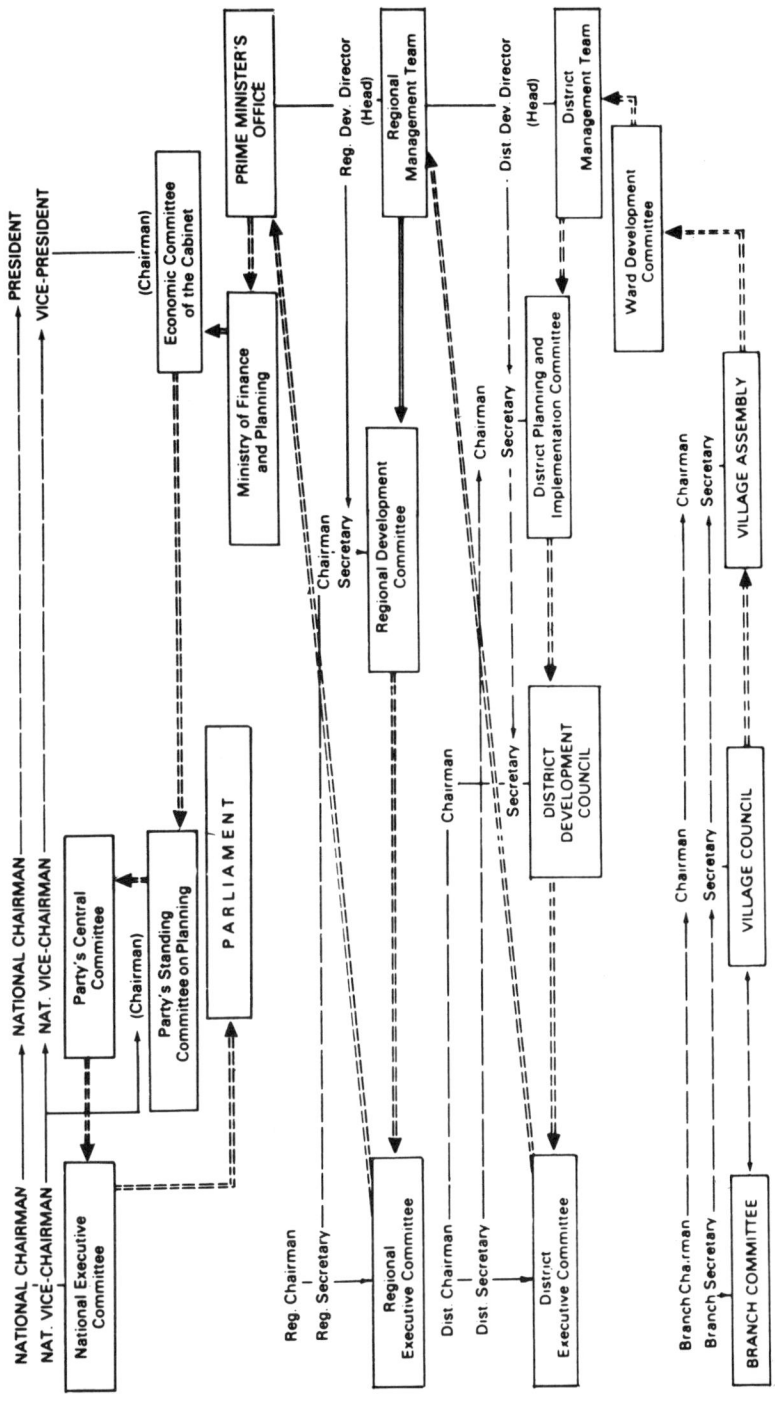

Source: Maeda (1983)
Figure 3b: *Principles for the Planning and Monitoring. Process involving Government and Party before 1984*

Manpower constraints are also serious. Although the civil service in Tanzania has expanded "at an astronomical rate" which "cannot be explained solely as a result of the expansion of the government services and activities",[10] a shortage of engineers, accountants and craftsmen has prevailed in the water sector throughout the last 15 years (SIDA/PMO, 1983, Ch. 3.4). Only Sweden has assisted in manpower development through the funding of a school for technicians in Dar es Salaam and by providing funds for the training of more than 150 water engineers in India. As the case studies indicate, other donors have put only limited resources into training (mostly on-the-job). To a large extent they have preferred to provide technical assistance personnel to compensate for the local manpower shortages.

The manpower problems are aggravated by the rapid declines in real incomes ever since the mid-70s (Stewart, 1986). Work motivation has decreased markedly as a result. In addition the "civil service is increasingly riddled by corruption and embezzlement of public funds" (Mukandala, 1983). Mukandala found, for example, that un-accounted-for cash expenditures in the public sector increased from 6 to 434 mill. Tsh. from 1973 to 1980. As the case studies will show, several donors have counteracted by bringing increasingly larger shares of the project activities under direct donor control.

Organizational problems are also considerable. The organizational structure at central level has been in a constant state of flux ever since the first Ministry of Water was established in 1971.[11] However, one key feature of the re-organizations has been that water activities have always belonged to a Ministry with responsibility for other major activities: energy/power (since 1974); minerals (from 1974 to 1981 and again in 1984); and now (since 1985) lands, housing and urban development. As Tanzania's economic crisis has deepened, the importance of both the energy and minerals sectors has grown disproportionately. The performance of both has a direct impact on the country's crucial foreign exchange situation. Furthermore, the capital funds controlled by the Ministry for energy and minerals sectors are far bigger than those for the rural water sector. In the early 1980s the latter only made up 10–15 per cent of the total capital budget of the Ministry. The budget for the urban water sector was of the same magnitude

---

[10] Civil service employment rose from 101,000 to 295,000 between 1972 and 1980 (Mukandala, 1983).

[11] There have been four major reshuffles from 1971 to 1985, namely in 1974, 1981, 1984 and 1985.

(URT, 1980). It is therefore not surprising that the rural water sector problems get very limited top-level attention at the Ministry.[12]

The final major institutional problem concerns the lack of intra- and intersectoral coordination. As explained in relation to Figure 3a, the central Ministry has slowly lost its coordinating powers and ability to provide technical backup to the regions. Thus, a report on the organizational problems of the rural water sector concluded that "after decentralization, there are 20 RWEs constructing rural water supplies as more or less independent organizations" (AIB, 1979, 33). Likewise the WMPCU – with its particular place in the organizational structure (Figure 2a) – has not been able to coordinate the long-term sector planning. It should be added that donors have increasingly preferred to provide technical assistance to the regions only. Thus the support to strengthen the technical and administrative capacity at central level has diminished significantly since the mid-70s, when SIDA started to reduce its technical assistance to the Ministry. The local support function of national headquarters has been ignored. This has been a general trend in other sectors as well during decentralization (Hyden, 1983, 94).

Coordination between donors and recipient authorities has also been a problem. During the early 1980s attempts to "establish a regular pattern for mutual consultation and coordination of water development strategies" between the Ministry and various donors failed. After initial enthusiasm – notably expressed by the World Bank – some donors got cold feet. Several of them resented the World Bank trying to take on the lead agency role. Others felt that donors might gang up against the Ministry if regular consultations within the donor group were to be held. They argued that although coordination was needed it should be the responsibility of the Ministry, not the donor group. And some donors feared the restrictions and standardization which coordination would be likely to entail, and which might affect their commercial or professional interests. So the attempt to establish a Tanzanian Water Development Coordination Board died as suddenly as it had appeared.[13]

With respect to inter-sectoral coordination, Figure 3a clearly shows that this involves numerous bodies: The Party, the Prime Minister's

---

[12]. Confirmed in interviews with high ranking Ministry staff.
[13] Quote from resolution 4, Morogoro Water Conference, 1980. Remaining information from files at WMPCU, Dar es Salaam; and interviews with donor staff, January 1985.

Office, the Ministry of Finance, and the technical ministries. This increases the complexity and diminishes the power of any one agency to enforce and direct both short and long-term planning and subsequent implementation.

In 1980 the Principal Secretary in the Ministry gave the following summary of the rural water sector coordination problems, a description which still holds today:

> ... the situation appears to be one of non-coordinated support and non-coordinated machinery to receive this support. And the irony of this picture is that all efforts are made with good will and with the best of intentions. (Mwapachu, 1980, 11)

2.4. Policy Problems

Policy-making for the rural water sector involves both the Party and the central ministries, with the "big" decisions being made by the former. Thus it was the Party which provided the tremendous push for improved water supply following the Arusha declaration in 1967. During a period of six years the goal for the rural water sector was changed three times. In the Second Five Year Plan (issued in 1969) water to all was envisaged within 40 years. Two years later (1971) the Party declared that this goal (safe water within 400 meters) should be reached within 20 years. And in 1975 – at the peak of the villagization campaign – the Party declared that by 1981 all villages should be provided with one source of good water within a reasonable distance. Compare this to the pre-69 period when rural water supplies were not even mentioned in the social service sector reviews (URT, 1968, 45–47).

Another important policy statement was also made in 1971. It concerned participation. Ever since the influential Mwongozo Party Guidelines were issued, both the Party and the government have reiterated the importance of participation. The guidelines, still in force, state:

> If development is to benefit people, the people must participate in considering, planning and implementing their development plans ... The duty of our Party is to ensure that the leaders and experts implement the plans that have been agreed upon by the people themselves. (TANU, 1971, para 28)

During the early 1970s few donors had their own policies on participation but presumably they were favourably disposed towards the Tanzanian emphasis on this issue.[14] More recently, participation has become a buzz phrase in the donor community as well (see chapters 3.5 and 10.4)

Tanzania's participation policy is based on a self-reliance ideology. However, in 1970 the Party abolished user cash contributions to the construction, operation and maintenance of rural water supplies (Boesen, 1986; Warner, 1970). The 1971 policy on water to all by 1991 reaffirmed the principle of water as a free public good. It is based on a welfare ideology. This conflict between a self-reliance and a welfare ideology has never been resolved. In practice the Party, the government and the donors have all followed a welfare approach to rural water sector development. For the 1971 policy, promising "free" public services to everyone within a short time, is popular. It corresponded then and still corresponds today with the priorities of most villages for government support, as extensive surveys show (BRALUP/CDR, 1982, Ch. 4). It assures an important role for various government departments and justifies their claim for increased resources in sector development. And when the improvement of water supplies is no longer restricted by local willingness and ability to pay, it can be provided by the State and it can be done according to need. Finally, this policy provides politicians with a valuable political resource. The political leadership has used water schemes as carrots to encourage peasants to start Ujamaa villages when this was still done voluntarily. And by 1974 an improved water supply was one of the benefits that peasants were promised in exchange for moving into village settlements. Local politicians and bureaucrats also found control of the funds for new water schemes useful. It became a valuable chip in patron-client bargaining in which members of Parliament have been especially active (Samoff, 1974, 84; Finucane, 1974, 146).

Hyden (1979) has called this style of policy-making "We-must-run-while-others-walk". It often leads to conflicts between means and ends which are left unsolved until the implementation stage. Thus bold policy statements may very well be reflected in the long and medium range (5-year) plans, but the actual resource allocations are fixed in the yearly budgets and they have been much smaller. Unfortunately

---

[14] Participation has been a fairly consistent theme in the minister's yearly speech to the National Assembly (see for instance Ministry of Water, Power and Electricity (1972) and Ministry of Water and Energy (1983)).

Table 2. *Resource Allocations to Rural Water Supplies in Tanzania, 1964–1984*

	unit	64/65–68/69	69/70–73/74	74/75–78/79	79/80–83/84
Capital expenditure					
Donor funds	m.Tsh	26	141	471[a]	649
Total funds	m.Tsh	39	193	617	1032
Total funds in 1974/75 prices[b]	m.Tsh	n.a.	n.a.	496	428
Donor funds/total funds	%	67	73	76	63
Total funds/total central & regional government capital budget	%	n.a.	n.a.	2.6	2.2[c]
Recurrent expenditure[d]					
Total regional funds	m.Tsh	n.a.	n.a.	286	763
Total regional funds in 1974/75 prices[e]	m.Tsh	n.a.	n.a.	222	238
Total regional funds/ total regional gov. recurrent budget	%	n.a.	n.a.	4.4	5.7

a Estimated for 1975/76
b Assumed that 60 per cent of donor funds spent outside Tanzania (see Ministry of Water *et al.*, 1984, 4.23; and Hordijk *et al.*, 1982, 72). These funds subject to inflation in the international price index for dollars. Remaining funds subject to local inflation after the retail price index. Indexes in Schluter (1982).
c 1979/80–1982/83.
d Rural water supply recurrent expenditure from central government budget cannot be separately identified. Figures indicate regional funds only.
e Deflated with retail price index.
Source: Therkildsen (1986).

much discussion on Tanzania's basic needs policies ignores this important fact. It is equally unfortunate if it is ignored by planners.

Table 2 is based on the yearly budgets. It shows that rural water supply development allocations have been a relatively small and declining part of total development allocations since 1974/75. During the same period there has also been a fall in real terms. These declines have been fairly consistent year by year since 1975/76. Donor contributions to rural water supply capital expenditures have also declined steadily since 1975/76 both in real terms and as a share of total funds.[15]

It is important to note that donors have their own "we-must-run" policies. They tend to prefer large projects to small ones; projects that can be implemented fast to slow-moving ones; and those that produce quantifiable results (Korten, 1980). Some of the consequences for planning and implementation of these biases are discussed in Chapter 3 and in the case studies (Ch. 4 to 8).

Another important problem is the *ambiguity of policies*. They are often unclear and non-operational and this affects planning and implementation. The priority criteria for selecting candidate villages for new schemes, for example, are variously based on "need", "development potential" or "least cost" (BRALUP/CDR, 1982, Ch. 12). Operational definitions of these concepts are, however, not given. Another example concerns operation and maintenance policies. A growing number of schemes went out of operation from the mid-1970s onwards, due – in part – to lack of recurrent finances. Several donors have put pressure on the Tanzanian authorities to introduce a cost recovery system based on direct user payment. In 1982 the principle of water as a "free" public service was challenged by the Party Chairman during the yearly Party conference: "Responsibility for looking after facilities must clearly be theirs (e.g. the users). The Government cannot finance the maintenance and repair work."[16] Does this statement from the highest authoritative source qualify as a new policy? Apparently not quite. By the end of 1985 it had not yet been operationalized although several donors pushed hard to get a policy clarification. The technical ministry (MAJI) and the ministry responsible for rural development (PMO) disagreed as to who should prepare a policy paper to the Economic Committee of the Cabinet. And once a paper was produced, the cabinet refused to make a decision.

But there are several other reasons for the ambiguity of policies. The Party monopoly on major policy-making does not encourage government ministries to take too active a role. Moris (1978) has noted that before policies have been announced, they are "sensitive" and "confidential", and not suitable for discussion. But once announced they become "sacrosanct". Thus, when only the means to reach the ends – but not the ends themselves – can be subjected to analysis, it is difficult or impossible to formulate clear and operational policies. It is a contributing factor that the institutional capacity for policy analysis is often weak, especially in the Party, but also in the ministries. The Party

---

[15] A peak in resource investments occurred in 1975/75, due to the extensive villigisation that took place around that time.

[16] Statement by Julius Nyerere as quoted from *Daily News*, October 21, 1982.

has not, for example, conducted any policy or performance review of water sector activities prior to 1984 (Mattoke, 1984).

Donors have their own policy problems. Development assistance fashions change over time. The major trend setters like the World Bank and the IMF often tend to influence bilateral donors. The bilaterals themselves sometimes change aid policies to fit in with domestic policy requirements and commercial interests. A donor may also introduce approaches and experiences gained in one recipient country into another one. In short, development assistance approaches and priorities are continuously adjusted often for reasons unrelated to conditions in the specific recipient country and region. Obviously this adds to the ambiguity and complexity of planning and implementing government projects. It also adds to recipient frustration. At a recent top-level meeting between a donor and the Ministry of Water, the Principal Secretary at one point asked "What, then, is your policy this year?"

This raises a final, crucial question. What are the real political bureaucratic interests in rural water supply? Are they mainly related to the benefits that politicians and bureaucrats obtain from controlling the allocation of new schemes as already discussed? Or is there a constituency with an interest in securing adequate resources for the maintenance of such schemes? What is the donor long-term interest in rural water supplies? Donor plans and donor implementation must build on genuine recipient interests if they are to have any significant long-term impact. Planning and implementation do not occur in a political vacuum.

2.5. The International Drinking Water Supply and Sanitation Decade

Experiences from Tanzania are of particular relevance in the context of the International Drinking Water Supply and Sanitation Decade (1981–1990). This Decade was declared by the UN General Assembly in 1979, following preparatory meetings on housing (Vancouver, 1976), water (Mar del Plata, 1977), and primary health care (Alma Ata, 1978). Biswas (1981) has described the water supply situation in developing countries around that time. The declared target – water to all by 1990 – was indeed an ambitious one, given that the average coverage in the rural areas of developing countries was around 32 per cent at the start of the Decade (see Table 3).

Tanzania may be considered a "pilot country" for Decade activities

Table 3. *Rural Water Supply Baseline Data. Global Surveys, 1970–1985*

	Global[a]			26 African countries			Tanzania		
	1970	1980	1983	1970	1980	1983	1970	1980	1985
Coverage									
Total rural pop. (mill.)	1210	1490	1610	215	263	300			
Rural pop. served (%)	13	32	36	13	22	31	10	33	40
Decade targets									
Total coverage (≠ countries)			13			2		yes	
50–95% coverage (≠ countries)[b]			66			21			
Less than 50% coverage			15			3			
Decade plans prepared or being prepared (≠ countries)			76			22		yes	
Capital expenditure									
Donor funds (% of total funds)	na	na	na	na	na	70	76	63	

Notes: a The global data exclude China. In 1983 a total of 94 developing countries were covered by the survey
       b May include countries without targets, see WHO (1986a, table A.2).
Sources: Therkildsen (1986): All Tanzanian data
         WHO (1986a)   : All remaining coverage data
         WHO (1986b)   : All remaining global and African data

for several reasons. Table 3 shows that many developing countries – as Tanzania did 15 years ago – have now established specific targets for rural water supply development. The table also shows that most developing countries (e.g. 22 out of 26 African countries) have produced, or are in the process of producing, Decade plans. Tanzania has considerable experience in this respect as Chapter 2.4 indicates. Finally, Table 3 shows how donors are a major source of funds for rural water supply development in Tanzania. Globally more than 100 inter-

Table 4. *Major Decade Constraints in developing countries. Global Survey, 1983.*

Funding limitations*	Import restrictions	Lack of planning and design criteria*
Operation and maintenance*	Logistics	Inappropriate technology*
Inappropriate institutional framework*	Insufficient health education effort	Inadequate or outmoded legal framework*
Inadequate cost recovery framework*	Intermittent water services	Insufficient knowledge of water resources
Lack of professional staff*	Non-involvement of communities*	Inadequate water resources
Lack of subprofessional staff*	Lack of definite government policy*	

Note: The constraints are ranked by column in descending order from left to right. The ranking is based on WHO (1986b), but adjusted for differences in response rates.
*Discussed in this report.

national aid organizations are involved in Decade activities[17] (WHO, 1985).

In Tanzania, donor assistance has gradually switched from programme to project aid over the last 15 years, and this has influenced water sector development in many ways. A similar switch can be observed in many other countries (Morss, 1984). This, too, makes it relevant to study in Tanzanian experience.

But it is more than the emphasis on targets, plans and external support that makes Tanzania a relevant case to study. Table 4 shows the major constraints to Decade development identified in a WHO-survey of 94 developing countries in 1983. When allowing for the ambiguities inherent in such global surveys, the list contains most of the issues dealt with in this and subsequent chapters on the rural water sector problems in Tanzania.

One final point is appropriate at this stage. The Decade documents emphasize the interrelationship between water supply, sanitation and

---

[17] This includes bilateral, multilateral and foreign non-governmental organizations. It excludes all domestic organizations.

health education, and their impact on health. In Tanzania there has not yet been much experience with the planning and implementation of such integrated activities.[18] Sanitation and health education activities are therefore not dealt with in this study. The focus is exclusively on the planning and implementation of rural water supplies.

2.6. The Research Questions and Methods

Based on the discussions above, this study focuses on two typical problems of donor involvement in rural development activities. They were observed during the author's work on the preparation of the Danish-funded RWMPs and during visits to various donor-funded rural water supply projects in Tanzania in the early 1980s. Similar problems have been noted by several other observers as well (Strachan, 1978; Lele, 1975; Rondinelli, 1982; Johnston and Clark, 1982).

One problem was that despite the fairly substantial resources and technical expertise being provided by donors for the preparation of RWMPs, they did not seem to be used much during the subsequent implementation. Most RWMPs were "collecting dust", according to the previous national coordinator of this planning exercise (Schønborg, 1983). The other problem concerned implementation. Following the completion of the medium- and long-term plans, some donors provided significant funds and technical assistance for the implementation of rural water schemes. This resulted in the construction of many water supplies but soon after completion many of them did not function properly, or were not used as intended by the villagers. The scheme benefits were difficult to sustain.

Obviously, part of the explanation for this is to be found in the steadily worsening conditions under which all rural development activities in Tanzania have suffered since the mid 1970s, (Stewart, 1986). The effects have been amplified by sector policies that were not adjusted to this deepening crisis (and about which donors asked few questions during the 1970s). Yet there was a considerable gap between what was planned and what was actually implemented. The rapid collapse of donor-implemented schemes was also striking – especially when donors attempted to hand over completed schemes to the Tanzanian organizations. Do, therefore, the particular donor planning and

---

[18] Among the five donors included in the case studies (Chapters 4 to 8) only Sweden is funding integrated activities, but only since mid-85.

implementation approaches in *themselves* contribute to the observed problems? Hence the specific purposes of this research:

(i) Which characteristics of the donor approach to the preparation of medium- and long-term plans may have contributed significantly to their use or non-use during subsequent implementation?

(ii) Which characteristics of the donor approaches to the preparation and implementation of these plans may have contributed significantly to the sustainability or non-sustainability of project benefits?

(iii) Which changes in approaches may improve the effectiveness of donor planning and implementation of rural development activities?

The analyses are based on studies of the rural water sector development in Tanzania in particular and of rural development in general. They cover the period from the early 1970s to the end of 1985. Five examples of donor involvement in rural water sector development have been selected as case studies: The Danes in Iringa, Mbeya and Ruvuma regions; the Finns in Mtwara and Lindi regions; the Dutch in Morogoro region; the Swedes in Mwanza, Mara and Kagera regions; and the World Bank in Mwanza region (see Figure 1).

Various factors were considered in the choice of case studies. Denmark, Finland, Holland and Sweden are four important donors that are still active in the rural water sector. The only other important presently active donors are Norway and UNICEF (WHO, 1985). Secondly, the patterns of involvement of the five donors have been different. The Finns and the Dutch started their involvement during the mid-1970s and put a major emphasis on the construction of new schemes. They are still involved. The Swedes funded the preparation of the RWMPs in 1975 but actual implementation was delayed until 1984 and has undergone several changes since then. The World Bank, reputed to be skilled planners and implementors, entered Mwanza region in 1977 with a relatively large project of short duration. By 1984 World Bank funding and technical assistance was stopped. The project collapsed. It was not linked to the Swedish-funded RWMP for that region. The Danes are the latest in the selected group to arrive in the sector. Their RWMP and their subsequent implementation are characterized by some emphasis on village participation in sector activities. Finally, availability of information has influenced this particular choice of donors. An overview of donor involvement in the rural water sector is given in Table 1.

The research is based on information from (i) interviews with a large number of Tanzanian civil servants, politicians and academicians, conducted during an uninterrupted stay in Tanzania from 1980 to 1983 and during visits in 1984 and 1985; (ii) interviews with technical assistance staff from consultancy companies and donors; (iii) unpublished reports, etc., from Tanzanian authorities and donors, and (iv) published records from these sources.

3. Planning and Implementation Approaches

Chapter 2 focused on some of the substantive problems in the rural water sector. They are all characterized by a considerable degree of complexity and uncertainty. This chapter focuses on some theoretical aspects of planning and implementation. They concern the processes and methods by which donor funded projects and programmes are designed and executed. They illustrate a range of possible approaches which are of particular relevance to the analyses and interpretations of the five case studies which follow in Part Three.

Together, Chapters 2 and 3 set the framework for the analyses and discussions of some main themes of this study: the relationships within the complex and uncertain context of the rural water sector; the various planning and implementation approaches used by donors; and the appropriateness of plans and sustainability of benefits resulting from donor assistance to rural water supplies.

3.1. Approaches and Aid Policies

In their well-known book on planning of rural water supplies in developing countries, Cairncross *et al.* (1980,1) categorically state that "planning will have a major part to play in rectifying the unsatisfactory conditions of many rural communities". Few would object to the statement at this level of generality. Controversy arises when the wide variety of different planning and implementation approaches are considered.

At the one extreme, comprehensive, long-range and detailed pre-implementation planning is advocated. This is the approach that is often advocated in textbooks, the gist of which is illustrated by the quote below (Shaner, 1979,24). The book is based on experiences in Ethiopia and the target audience is graduate students from developing countries:

> A project, ideally, consists of an optimum set of investment-oriented actions, based on comprehensive and coherent sector planning, by means of

which a defined combination of human and material resources is expected to cause a determined amount of economic and social development. The components of a project must be precisely defined as to character, location and time. Both the resources required – in the form of finance, materials, and manpower – and the generated benefits – such as cost savings, increased production, and institutional development – are estimated in advance. Costs and benefits are calculated in financial and economic terms or defined (if quantification is not possible) with sufficient precision to permit a reasoned judgement to be made as to the optimum set of actions.

Some practitioners are less ambitious and perhaps more realistic. DANIDA (1985), for example, has issued a set of guidelines for project appraisal and planning that represent a more flexible and less demanding approach. The guidelines put some emphasis on pre-implementation planning but do not require detailed specifications of medium and long-term activities.

Korten represents a more radical view. He claims that the key to rural development planning and implementation:

> . . .is not analytical methods, but organizational process; and the central methodological concern is not with the isolation of variables or the control of bureaucratic deviation from centrally designed blueprints, but with effectively engaging the necessary participation of system members in contributing to the . . . knowledge of the system. (quoted from Rondinelli, 1983, 130).

With respect to rural water sector planning it is the first mentioned extreme, or derivations of it, that are typically advocated (Cairncross et al., 1980; Widstrand, 1980, 108; Lium and Skofteland, 1983). An explicit example is provided by the World Bank (Grover, 1983). Its guidelines list 266 issues that should be analysed in a standard pre-feasibility study of water supply and sanitation projects. These issues should then be elaborated on in subsequent pre-implementation planning. It is not accidental that the case study presented to exemplify the use of these guidelines is from the imaginary Republic of Optima. Indeed, it is this type of approach which is generally used in most donor assisted rural development activities, despite much rhetoric about flexible bottom-up planning and implementation (Johnston and Clark, 1982).

It is interesting to note how the type of planning and implementation approach proposed in the textbook quoted above has been regarded as the ideal over the years, while the aid policies of many donors have changed significantly during the 1970s (Rondinelli, 1983, Ch. 1). The earlier predominant concern with promoting rapid growth in gross national product through capital intensive industrialization, export promotion and large-scale construction of physical infrastructure, has partly shifted towards a focus on stimulating internal demand, expanding peasant agricultural production, and improving basic needs of rural people. Despite these changes towards poverty-alleviation in rural areas the planning and implementation approaches preferred by donors have remained those used during the earlier phases. They originate from Anglo-American physical planning. These approaches are so widespread and commonly accepted that it is important to point out that a number of important variations exist.

3.2. Functional Versus Normative Planning

Along the functional-normative dimension there is a basic difference in the extent to which planners regard ends as given (Faludi, 1973a, Ch. 9). In functional planning, planners accept the substantive goals – whether they are set by a higher authority or a donor, by a political process, or by the planners themselves. Only various means to reach the ends are subjected to analyses. In normative planning, planners scrutinize both means and ends.

The particular mode used depends on two main factors (Faludi, 1973a, 177). One concerns the relative autonomy of the planning agency. The other concerns the bureaucratic-political roles of the planners. If a planning agency is relatively autonomous and politicized, the scope for normative planning is substantial. The agency then scrutinizes both means and ends and advocates changes despite the inevitable conflicts and opposition that this may cause. If, on the other hand, the planning agency has only a small degree of relative autonomy, and only a bureaucratic or technocratic role of the planners is acceptable, functional planning is likely to be practised. The placement of the planning agency in the organizational hierarchy is obviously an important factor in relation to its relative autonomy.

In the context of rural development planning based on foreign assistance, and thus involving two governments, these considerations are helpful. It is clear that functional planning of donor assistance may cause irrational or undesirable results, because the ends as stipulated by the recipient or the donor may be infeasible, unrealistic, unjust, etc.[1]

The case studies in Chapters 4 to 8 provide vivid illustrations of planners that unquestioningly base their plans on questionable policies. It is equally clear that normative planning of donor assistance can be contentious. It may suppress recipient policy making responsibilities and increase its external dependency. Conversely it may turn the donor or its planning consultants into *de facto* policy makers whom "policy-endurers" (for example villagers) cannot hold accountable at all. The case studies illustrate that the donors involved in the rural water sector have become increasingly interventionistic since 1980, although some prefer to use the more diplomatic phrase "policy dialogue" for this. Whichever the case, the choice of approach along the functional-normative dimension is a key one – and inherently political.

So far conventional planning theory is useful. But both the functional and the normative mode of planning are based on the assumption that a certain degree of consensus on objectives exists – or can be made to exist. The possibility that there is no real consensus is ignored in conventional theory, as demonstrated by Faludi (1973a, Ch. 9).[2] Yet there may be only limited consensus on the objectives of a rural development activity despite a formal agreement between a recipient and a donor (Moris, 1981); the consensus may only be temporary and subject to change depending on the power relationship between the two parties and the beneficiaries (Strachan, 1978; Grindle, 1980); the objectives themselves may be unclear or ill-defined (Rondinelli, 1983); or they may not reflect the needs of the intended beneficiaries (Korten, 1980).

The issue of consensus on rural water sector objectives is a crucial one for planning and implementation approaches and is a theme that is pursued several times in the rest of this study.

3.3. Blueprint versus Process Planning

The central feature of blueprint planning, according to Faludi (1973a, Ch. 7), is the *plan*. It "will formally consist of one or more goal statements, ... specific policies, programs, and projects, all ... related to sets of priorities, standards, investment needs and financial arrangements". These specifications are made for considerable time

---

[1] From the planners' point of view.
[2] Faludi mentions a number of ways in which objectives can be set. He does not distinguish between formal and *de facto* acceptance of these objectives.

spans and are based on analyses performed in *advance* of implementation. Pre-implementation information needs are therefore considerable. Monitoring and evaluation of the implementation itself is done to obtain information on the basis of which deviations from plans can be controlled.

The process mode of planning, on the other hand, is an approach whereby activities "are adopted during implementation as and when incoming information requires such changes". The plan document itself becomes far less significant than in the blueprint mode. "Information and feedback impinge directly on action, providing signals that lead to incremental adjustments" of both means and ends. The planning horizon is not necessarily a short one. Continuous adjustments may be made within a long-term framework provided by a strategy or a set of policies which are themselves subject to review as new information is forthcoming. Thus "process planning operates simultaneously on several time horizons" (Faludi, 1973a, 132).

There are three factors that influence – or ought to influence – the choice of planning mode along this dimension according to Faludi (1973a, Ch. 7). One concerns the degree of context stability. In blueprint planning it is assumed that the environment in which activities are implemented is so stable and predictable that stated objectives can be reached with certainty. If this assumption cannot be made, a process mode of planning is more appropriate. The second key factor concerns control. Blueprint planning is based on the assumption that plan compliance can be ascertained because the environment is stable and implementation activities can be controlled by one or a few agencies. Conversely, if multiple decision makers exist, agencies must strike bargains, adjust to each other, and respond to opportunities as they arise. This requires seeking information through a network of channels and responding with flexibility, which is what is implied by process planning. The third factor concerns information time-lags. When a long space of time elapses from an agency receiving information from the field until it formulates new plans or adjusts activities, the possibility for process planning is restricted. This internal time-lag is obviously a real problem in many donor and recipient bureaucracies. But external time-lags are also important to consider. They concern the time it takes to obtain information about the outcomes or results of activities. In many rural development activities such outcomes are often difficult to measure and slow to appear.[3] Everything else being equal, this tends to encourage blueprint planning, according to Faludi (1973a).

In the newer literature on rural development planning and implementation, the information issue is intensively debated (Korten, 1980; Rondinelli, 1983; Gow and Morss, 1985). It is pointed out that information obtained by monitoring and evaluation tends to be treated differently in the two modes. In the blueprint mode, information from monitoring of implementation is primarily used to *correct errors* (e.g. correct deviances between planned and actual activities by changes in the latter). In the process mode such information is used to *learn from errors* (e.g. deviances between planned and actual activities may lead to changes in both).

Control is a central issue here. Correction of errors presupposes that the power to do so exists. Implicitly, many donors and recipients assume that blueprint plans can be implemented because one or a few authoritative and powerful decision makers exist with the means and incentives to enforce compliance. This view may be based on wishful thinking or on faulty analyses of the nature of the state in developing countries (Hyden, 1983; Clapham, 1985). Whichever the case, blueprint planning continues to dominate when donors are involved in rural development activities. The issue of control and multiple decision-makers is therefore an important theme in this study.

3.4. Rational – Comprehensive versus Disjointed Incremental Planning

This is the third major dimension of planning according to Faludi (1973a). Among other things it concerns the types and extent of analyses on the basis of which plans are prepared. The issue of the political role of planning is also involved.

The rational-comprehensive approach to planning is aimed at identifying the most appropriate means to reach desired ends. Ideally, this requires a comprehensive search for alternatives to meet ends, and a rational analysis of each alternative to identify the best one – taking into account every important relevant factor. The methods used are from the tool kit of cost-benefit, systems analysis, Pert, input-output analyses, and so on.

The disjointed-incremental planning mode builds on more piecemeal analyses. It is the act of "muddling through" (Lindblom,

---

[3] However, complete information about the long-term results of activities is not always a major constraint for process planning, as Faludi (1973a, 141) implies. Decisions can often be made on the basis of short-term or proxy outcomes.

1959 and 1979). Means-ends analyses are often limited in number and in scope. Only a few alternatives are considered. The "best" alternative is regarded as the one on which there is agreement. Frequently this alternative does not differ much from present praxis; it is only incrementally different. Finally, analyses are often drastically reduced: important alternatives are neglected; significant outcomes ignored; or major factors not considered. But analyses of means-ends are constantly made, resulting in incremental decisions geared more to the alleviation of present problems than attacking future ones.

Much planning literature presents the rational-comprehensive mode of planning as the ideal one (Faludi, 1973a). Although it cannot be met it is an ideal that must be strived for. This point of view was strongly disputed by Lindblom (1959). Criticism has also been strong among writers on rural development planning (Rondinelli, 1976; Korten, 1980; Webber, 1983).

Several factors influence the choice of approach along this dimension. The basic but implicit assumption is perhaps political. Much discussion about the comprehensive-disjointed dimension concerns the desirability of central planning and its concommitant need for control versus decentralized planning by many independent decision makers (the market substitute). However, in the context of rural water supply planning, the factors of predictability, knowledge and information are more directly relevant. Rational-comprehensive planning is based on major assumptions on all three issues: (i) that future development concerning key factors is predictable so that long-term analyses of them are both desirable and possible; (ii) that sufficient knowledge exists to analyse means-ends relationships in a comprehensive way; and (iii) that adequate information is available – or can be made available at acceptable cost. The disjointed-incremental mode of planning is not based on these assumptions to the same extent.

Therefore unpredictability, lack of knowledge, and lack of operational information about key rural water sector activities constitute a third main theme in this study.

Since Faludi presented his state-of-the-art review of planning theory, many new ideas on process planning in the rural development context have emerged. Participation and institutional development are important issues in this new development of theory – they were hardly touched on by Faludi fifteen years ago. These issues, that encompass both planning and implementation, are discussed separately below.

3.5. Participatory versus Non-Participatory Planning and Implementation

The fundamental difference between a participatory and a non-participatory approach lies in the perception of the development process itself.[4] Thus participation by the intended beneficiaries in rural development activities may serve several purposes. It may help to secure user *acceptance* of government funded activities; it may help to solicit user *knowledge* about local-specific conditions needed for planning and implementation; it may help to obtain user *resources* in cash or kind to match external assistance; and it may help to increase user *power* and control over development activities both in the planning and implementation stages.

Even this short description implies that there are many forms of participation. Arnstein (1971) has identified eight of them, ranging from outright manipulation to real user control, but the fundamental distinction is between participation as a means and as an end (Oakley and Marsden, 1984, Chapter 2). Regarded as a means, participation aims at "mobilizing" the intended beneficiaries to take part in activities for which the contents have basically been determined from the outside. In this case user acceptance, knowledge and resource commitments are solicited. Participation regarded as an end aims at "empowering" the intended beneficiaries so that they may share in the control of resources, organize to control their means of livelihood, and take action to bring about structural changes that increase their power (see also Smith *et al.*, 1980, 17).

Non-participatory approaches to planning and implementation originate in the view that the key to rural development originates entirely outside rural areas – in the agents of the state and the donors supporting it. Implicitly many donors hold this view, as Chambers (1983, 115–32) has argued.[5] This view also dominates in Tanzania and is based on the nature of the post-colonial State. Despite policy statements to the contrary, it has a clear paternalistic bias related to the existence or emergence of a bureaucratic ruling class whose members attempt to expand their control over society in order to capture the surplus necessary for their reproduction. The dominant ideology of

---

[4] The vast literature on the development of the concept of participation and its practical applications are not dealt with here. See Oakley and Marsden (1984) for a recent review.

[5] A similar view is also prominent in much academic writing on the subject, as implied in the following Chapter headings: "Managing Participation" (Bryant and White, 1982, Ch. 10); "Organizing the Rural Poor" (Johnston and Clark, 1982, 164).

this class is modernization with which it justifies a policy of intervention and coercion in the rural areas (Stein, 1985). Rural development activities are planned, implemented and maintained *by* the state *for* the rural people. It is development from above (Boesen *et al.*, 1977).

The non-participatory approach is based on a welfare notion of development. The State is the benevolent provider of services to passive beneficiaries. It is assumed that users will accept the services provided; that the State, the donor and their agencies have sufficient knowledge to plan and implement village level activities; that user resource commitments are not important (or not important enough to solicit); and that the State and donor operate in the best interests of the beneficiaries, whose influence on development activities directly affecting them is therefore neither necessary nor desirable.

Such assumptions are not made in the participatory approach. User involvement is actively sought. In the context of rural water supplies it is thought to be essential for scheme sustainability (Miller, 1979). However, participation is very difficult to operationalize (Gow and van Sant, 1983 and 1985; Cohen and Uphoff, 1980; Chambers, 1983). No standardized models exist, nor can they be developed because of the context-specific character of the concept. However, recent surveys indicate that certain participatory methods are in general use although their specific applications vary a lot (Oakley and Marsden, 1984, Ch. 3). These methods concern four main issues.[6] First, *group formation* and the use of *paraprofessionals*: The roles of women in village level groups and as village-based paraprofessionals are key issues in the context of rural water supply development. A substantial literature on this issue has been reviewed by Wijk-Sijbesma (1985). Secondly, the role of the *outside agents* is important. They may help to facilitate the access of villagers to development activities, as has customarily been the task of the community development officers in many previous English colonies such as Tanzania. They may also assist villagers to build up an organizational base at grassroot level which would empower users to take a more active and assessive part in activities. The third methodological consideration concerns *procedures and structures*. Government and donor agencies must change and adjust planning and implementation procedures to make participation possible (whether "mobilizing" or "empowering"). Structural changes may also be required, such as a more decentralized decision-making. *Training* is the fourth key factor.

---

[6] Oakley and Marsden (1985) do not discuss paraprofessionals, procedures and structures.

It is important if user groups and paraprofessionals are to play active roles in development activities. The type of pedagogy and training – whether focusing only on skills or on "conscientisation" as well – is crucial in shaping the participation process.

The concepts of mobilization and empowerization underly this brief methodological discussion. Is the former possible without the latter? What are the limits of the former? How can the latter form be brought about?[7] These are key issues in any rural development programme that aims at participation. Unfortunately, the case studies presented here do not provide much useful information in this respect. But they do illustrate that participation is still an elusive concept in the context of donor assisted rural water projects.

3.6. Bypassing versus Institutional Development in Planning and Implementation

This dimension concerns the organizational arrangements and management procedures used by donors in the planning and implementation of projects. The central issue is the relation of the donor project to recipient government institutions.

Donor projects that bypass recipient organizations normally share a number of common features. Typically they are managed by Project Management Units (PMU) operating more or less independently of recipient organizations. PMUs normally substitute a range of activities that are already the responsibility of the existing government structure. These units are partly staffed by technical assistance staff. Compared to the indigenous administration, the PMUs enjoy a significantly greater independence and operate with a shorter chain of command. They frequently operate according to their own procedures, reimbursement schedules and objectives. Moreover, the PMUs usually have access to greater financial resources and managerial and technical expertise than the local organizations (Lele, 1975; Honadle *et al.*, 1983b). The PMU has been a common feature of donor assistance to the Tanzanian rural water sector (Ch. 4 to 8).

A PMU is, by its nature, temporary. It may cease to exist when

---

[7] Saul (1972, 25) argues, with reference to Tanzania, that mobilization and empowerization should occur concomitantly. "An (effective mass) base must be further mobilized, raised to consciousness, in essence *created* by effective leadership, *at precisely the same time* that its emergence is ensuring a continuing pressure on the leadership to play just such a creative role!" (Emphasis in original.) How this boot-strap operation is to take off is not discussed.

donor assistance to it stops. If continued, the PMU activities are transferred to an existing permanent institution, or the PMU is converted into a permanent institution operating compatibly with the remaining government structure.

The institutional development approaches to donor-assisted planning and implementation also have a number of features in common. One is that donor inputs are channelled through existing, permanent institutions. This normally implies an emphasis on programme rather than project support, although institutional development efforts are not incompatible with a project approach (Honadle and Rosengaard, 1983a). The other main feature of the institutional development approach is that the focus is not only on production but also on institutional development (Honadle et al., 1983b).

Despite its recent prominence, the focus on the institutional and management aspects of development assistance is not new. Rondinelli (1985) lists eight different approaches that have been tried out since the 1950s.[8] Each has focused on a different organizational level of the existing institutions and emphasized different problems. This has resulted in interventions aimed at changes in one or more of the following: (i) organizational structure; (ii) administrative processes; (iii) resource and input management; (iv) human resources and their training; and (v) the political, social, economic, and cultural context of development activities. However, the approach is not an established science which can provide the practitioner with a commonly accepted set of "tools". It is, in Rondinelli's words, "more an art than a science and, perhaps, more a craft than an art."

The role of the technical assistance staff in this type of activities is one of a "teacher" or "mobilizer" rather than one of a "performer" or "substitute" – to use the terminology of Honadle et al. (1983b). Priority is given to enhancing local skills; identifying and developing new skills; improving procedures; increasing beneficiary influence; and helping to influence other agencies whose cooperation or acceptance is important for reaching the intended objectives of the donor-assisted activities.

When bypassing is chosen it is related to perceived or real problems which are assumed to be easier to solve outside than inside existing recipient organizations. Typically, a PMU is established by a donor to

---

[8] They range from the technology transfer and community development approaches in the 1950s to the learning process and bureaucratic reorganization approaches in the 1980s.

exert a higher degree of *control* over planning and implementation activities. Such controls may aim at securing financial accountability and prevent misuse or embezzlement of donor resources. Increased donor control through PMUs may also aim at favouring donor suppliers of consultants, equipment and materials, thereby benefiting donor country employment and balance of payment. Sometimes control through a PMU is simply assumed to be needed to reach the intended target group if the recipient bereaucracy is deemed incapable or unwilling to do so.

Bypassing is sometimes also preferred when *experiments* with activities, which are not yet undertaken by existing organizations, are aimed at. Such experiments often require a large degree of initial independence (Paul, 1982; Rondinelli, 1983). This independence provides a PMU with the flexibility to experiment, take risks, make errors, and introduce changes which are needed to test new ideas. Rural development projects that experiment with new technologies, new delivery systems, or aim at broad socio-political objectives are often started up by independent project units.

Finally, bypassing is often the result of a strong emphasis on *efficient production*. Sometimes this is intended, as when a large, complex and unique product (e.g. a dam or a factory) is constructed. Efficient production can also deliberately be given a higher priority than the institutional development benefits that may result from the slower and more cumbersome production through existing recipient organizations.[9]

The main motive for the focus on institutional development is concern about the long-term sustainability of donor-assisted activities. It is assumed that the project approach to donor assistance and the concomitant bypassing of indigenous institutions may result in temporary infusions of resources, but that the benefits will cease soon after the donor support stops. Therefore institutional development, together with a focus on recurrent cost financing, are seen as necessary components of donor activities (Johnston and Clark, 1982).

---

[9] Unintended production orientation may also occur through goal displacement. Typically it happens when a project sets out to implement a number of activities that are expensive (at least initially) and whose outcomes are difficult to measure (training, structural and procedural changes, participation by beneficiaries, etc). The larger proportion of expenditure in a given project on such "software" components, the greater appears to be the need for ambitious production targets to justify the cost of the indirectly productive activities. The result is a contradiction in the project design from the very outset. Several case studies illustrate this.

Table 5: *Major Assumptions and Characteristics of Control-Oriented and Adaptive Planning and Implementation Approaches*

	Control oriented	Adaptive
Long-term objectivies	Commonly agreed, clear and reasonably consistent; often expressed in detailed production targets	Unclear and ambiguous; sometimes inconsistent; subject to negotiated change both with respect to short- and long-term objectives
Decision Making	Power to plan and implement rests with one or a few agencies in a fairly centralized decision-making hierarchy.	Power to plan and implement depends on multiple agencies and groups. Coalitions between them change.
	Within agencies decision making by planners—not implementors or researchers.	Within agencies decision often shared between planners, implementors and researchers.
Analyses	Environment is predictable. Uncertainty about cause-effect relations can be reduced through comprehensive analyses prior to implementation.	Environment often unpredictable. Uncertainty may be reduced through gradual learning process during implementation.
	Information is reliable.	Information is often unreliable and costly to obtain. Limited and selectively used during planning.
	Implementation information used to correct deviances between plans and actual activities.	Implementation information used to learn from errors leading to changes in both plan and activities.
Participation	Planned and controlled from above. Emphasis on mobilization or non-participation.	Evolves through learning process. Emphasis on mobilization and empowerization of beneficiaries.
Institutional capacity	Can be designed and implemented. Often technical assistance substitutes for perceived deficiencies. Bypassing typical. Often project-oriented.	Must evolve and grow. Size and implementation speed of external assistance adjusted to recipient institutional capacity. Technical asistance for capacity-building. Often programme-oriented.

The organizational framework for donor assistance to the rural water sector is an important theme in the case studies which follow.

3.7. Two Alternative Planning and Implementation Approaches

In theory the number of possible combinations along the five dimensions presented above is very large. However, a simplification of the myriad of combinations is possible. In practice, some modes are mutually exclusive (for instance, blueprint planning and participation as an end). Modes at the extremes of each dimension are rarely used in practice. The fundamental distinction between the various combinations is between control and adaptation.

In the former the major donor concern is that activities should be *controlled* through extensive and detailed pre-planning, through strong influence on recipient institutions (which are sometimes bypassed), and through a pre-determined and usually limited role of beneficiaries.

In the latter, the major concern is that planning and implementation should be *guided* by a long-term strategy, by experiences gained during implementation, by linking planning and implementation within recipient institutions, and through a continuous dialogue with the intended beneficiaries in all activities.

Table 5 indicates some main assumptions and characteristics of the two major approaches with respect to objectives, decision making, analyses, participation and institutional capacity. They are extracted from the previous discussions in this chapter and from Pfeffer (1981). These issues will be discussed again in the last part of this study, on the basis of the findings from the case studies.

Part Two
Case Studies

4. The Turn-Key "Approach": The Finns in Mtwara-Lindi Regions

The Mtwara-Lindi Rural Water Supply Project is the oldest continuously-supported water project in Tanzania; one of the first donor-assisted water projects with a clear regional focus; and the first rural water sector activity supported by Finland anywhere in the Third World. From a planning and implementation point of view the project has several interesting features:

- the donor and its consultants (Finnwater) have been involved in the two regions for 13 years without interruption;

- the consultants have acted as a turn-key contractor responsible for all aspects of the project from long-range water master-planning to subsequent construction, operation and maintenance of schemes (FINNIDA, 1984, 3);

- both the RWMPs and the subsequent implementation plans are based on the target of water for all by 1991 as expressed by the Tanzanian authorities (Finnwater, 1985a, 3); this has remained the target until 1984;

- the basic approach to implementation has not changed much from 1978 to 1983, but is now being drastically altered.

This turn-key approach – in which the consultant to the donor has controlled all activities – is briefly analysed below.

4.1. Description of the Project, 1972–1984

Finnish involvement in the Tanzanian rural water sector dates back to 1972. It started with a five-year preparation period costing 37 mill.Tsh. during which the RWMPs were produced. Then followed an

Table 6. *Finnish Funding and Output of the Mtwara-Lindi Projects, 1972–1987*

Project activities	Period	Total cost[a]	Source of funding[a]				People served[c] (×1000)	
			Fin-land	Tan-zania[b]	UK	UNI-CEF	piped	handpump
Project Preparation								
Feasibility study	1972–73	3.0			–	–	–	–
Housing Project	1973–74	3.0			–	–	–	–
Water Master Plan	1974–1977	31.0			–	–	–	–
Implementation								
Phase I	Jan. 78–March 80	33.9	30.0	3.9	–	–	130	120
Phase II	Apr. 80–Dec. 81	54.3	25,3	5.8	20.2	3.0	40	125
Phase III	Jan. 82–Dec. 84	59.8	32.0	4.8	5.6	17.4	–	175
(Phase IV)	(Jan. 85–Dec. 87)	(86.4)	(n.a.	n.a.	n.a.	n.a.)	(32)	(140)

a Mill. Tsh.
b Local funds for Finnwater-implemented activities. In addition GOT and Christian Council of Tanzania have invested some 64 mill. Tsh in rural water supplies in the two regions implemented through the RWEs (Finnwater, 1985b, 16).
c Estimates as given in project documents; actual population served may be significantly lower (see Ch. 4.1).
Sources: Finnwater (1980); FINNIDA (1984, App. 8, Table 1); Finnwater (1985a).

implementation period (1978–1984) divided into th
1985 around 148 mill.T.shs. had been used for i
through Finnwater in the two regions (see Table 6). T
in the construction of 1750 handpumped wells and 1
diesel pumps. By the end of 1984 Finnwater clair
schemes served almost 600,000 people – on the assur
schemes operate and are utilized as designed. In addition, schemes funded by other sources (local and foreign) should be included. The end result is that around 60 per cent of the population in both regions have access to an improved water supply (Finnwater, 1985b, Tables 9 and 10). The installed supply capacity in Mtwara and Lindi regions is now the highest in the country.

During the whole implementation period the Finnish funding has been fairly consistent. From 1980 it has been supplemented by funds from the United Kingdom and UNICEF. All donors have stuck to Finnwater as the consultant throughout. The consultants have acted as turn-key contractors. They have been responsible for "all details of the programme from engineering designs to actual construction of water supplies" (FINNIDA, 1984, 10). FINNIDA and the central, regional and district authorities (including MAJI) have only played a limited role in the whole exercise. Finnwater has simply provided ready-to-use water schemes to the villages – and has maintained them too. However, since mid-1983 some degree of village involvement has been introduced (see below).

The RWMPs, prepared from 1975 to 1977, were based on the 1991 goal. Up until 1984 implementation planning was also geared towards this goal. It is now being abolished. A revision of the old RWMPs is being made for the period 1984–2001, but based on estimates of available financial resources (Finnwater, 1985b, Ch. 10). The turn-key approach will also be abolished. These changes were initiated when a new Project Manager and a Training Officer arrived in early 1983. They were, furthermore, strongly encouraged by a very critical evaluation (FINNIDA, 1984) which pointed out the obvious sustainability problems involved in running a Finnish-funded "Water department" independent of local authorities and villages. There is no indication that the Ministry – to which the project is formally attached – played any prominent role in this change of direction.

The main aim of the new approach is that Finnwater should gradually hand over the responsibility for actual construction and maintenance to local authorities and to villages. When this will be accomplished is not specified in the project documents and remains uncer-

see Ch.4.8).¹ Another feature of the new approach is that health, education and sanitation activities will be initiated and that special efforts will be made to involve women (Finnwater, 1985a).

4.2. Planning

Preparation of the RWMPs was the stepping-stone for the Finnish-funded construction of water schemes. Finnwater worked out a 13-volume plan on the basis of a two-page general description of the scope of work. The plans were prepared prior to any Finnwater involvement in implementation. There is no indication that the Ministry or the local authorities were involved in any significant way in the plan preparations.

The plans contain a water resources inventory; a proposal for developing these resources in phases to meet the 1991 goal; and a shopping list of projects to be carried out immediately after the completion of the RWMPs.² A crash programme for providing each village with one reliable source of water by 1981 was also prepared. To reach the 1991 goal was estimated to cost 239 mill.T.shs. and 20 mill.T.shs. yearly in operating expenses by 1991 (Finnwater, 1977a). Most of the analyses contained in the RWMPs concern technical issues. There are, however, no analyses of major technical alternatives. And there are no analyses of the administrative, economic and social consequences of the technical proposals and implementation programmes included in the RWMPs.

Forty-five various Finnish expatriates and 84 Tanzanians were involved in the preparation of the RWMPs. The Finns stayed on average 16 months in the regions. Only one professional Tanzanian hydrologist counterpart was attached to Finnwater – for seven months. The remaining Tanzanians were either technicians (14 persons) or unskilled (70 persons). The RWMP exercise was basically a Finnwater operation without much Tanzanian input (see also Ch. 4.3).

The planning of the subsequent implementation phases has also been carried out by Finnwater. The evaluation team of 1984 regarded

---

[1] Finnwater has planned that previous construction rates should be kept in phase four (1985–87) but that the MAJI offices in the two regions should gradually take over this activity from Finnwater. FINNIDA is to provide finances for construction by both agencies.
[2] Chambers (1973) has pointed out that a shopping list at the end of plan documents is a general feature of much rural development planning.

these plans "more as a general programme guideline than a specific and complete description of services" (FINNIDA, 1984, 10). As was the case for the RWMPs, these implementation plans contain no analyses of alternative proposals (not even technical alternatives), and non-technical issues were generally ignored.[3]

One would perhaps expect that with so much consultancy assistance and such large funds involved, a set of objectives for the RWMPs and the implementation would have been formulated. Not so. "A formal statement of the project objectives has apparently not been made". Yet a "general understanding of the main objective" has emerged. It is to improve the rural water supplies to achieve improvements in general health and economic development (FINNIDA, 1984, 9). But only one objective was actually followed in the field until 1984. It was simply to provide improved water supplies at the lowest cost as fast as possible. This production orientation has significantly shaped many decisions in the implementation phase – regardless of the proposal in the RWMPs – as will be shown below.

4.3. Institutional Framework

Since the preparation of the RWMPs, Finnwater has operated largely independently of the Tanzanian authorities. All contracts have been signed with FINNIDA – although subject to approval by the Ministry. Both the RWMP exercise and the subsequent implementation have been designated as "national projects" which fall under Ministry jurisdiction (see Ch. 2.3). Consequently, the regional authorities have mostly watched events from the sidelines together with the villagers. This bypassing of local institutions has been almost complete up to the beginning of 1984. Now more formal links to the regional authorities are being established through a Project Steering Committee with regional representation.[4] Until recently, operations in the villages were also made without much more than cordial contacts with village governments (Ch. 4.9).

Basically, Finnwater has operated on its own. Although the Ministry is formally the counterpart institution, it has not taken an active

---

[3] From mid-1983 the non-technical aspects of the water activities have, however, gained increasing prominence.

[4] Finnwater was initially against the idea. The argument was that "national" projects like the Mtwara-Lindi one is the responsibility of the Ministry, not the regional authorities (see Ch. 2.3). The latter should therefore not be involved in decision-making.

part in shaping programmes and plans – nor in supervising their implementation. It has simply been too difficult to get Ministry staff to come to these peripheral regions for any prolonged stay. Even Ministry attendance at meetings in the regions has sometimes been a problem. Thus initiative and responsibility have been with Finnwater and its staff.

Three examples illustrate this. There has been no professional counterpart staff attached to the project during implementation (FINNIDA, 1984, 24). All positions requiring middle and high-level skills have been occupied by the consultants' own staff. This staff has consisted of 13 to 21 people during the implementation period. They operate strictly as "performers" (see Ch. 3.6). They control all aspects of the project from planning to maintenance of completed schemes. Lower-skilled (70 persons) and unskilled Tanzanian staff (200 persons) are employed directly by Finnwater and are not seconded from MAJI. This staff enjoys significantly better terms of services and endures stricter discipline than their colleagues in the government system. Secondly, Finnwater itself imported most equipment and materials directly from overseas. Mtwara has an excellent harbour acceptable to shipping lines. In this way the project has become independent of the supply situation in Tanzania. Finally, to speed up work in the field, villagers have until 1983 been paid to dig trenches (Tsh 1 per metre) and wells (Tsh 500 per well). Apparently this arrangement did not prove efficient enough. Tractor excavators were therefore increasingly used until this was stopped in 1985.

Isolating the project from the Tanzanian environment, right from the village level and upwards, has obviously had negative impacts on the capacity of local institutions to take over once the donor withdraws support. By 1985 neither villages nor MAJI were prepared for this, despite (or because of?) 13 years of continuous donor assistance.

4.4. Scheme Technology

The choices made in the RWMPs were not based on any systematic analyses of the pros and cons of different technologies. Instead, the planners appear to have based their decision on a mixture of a one-variable criterion and politics. Thus piped water supplies fed from ground water sources were given first priority because ground water generally is much cleaner than surface water (Finnwater, 1977b, 11). Also Tanzanian preference for piped supplies at that time undoubtedly influenced the plan proposal. Hand-pumped wells were only to be used in special cases (scattered, small settlements). Thus, the RWMPs envisage that 92 per cent of the population will be served with piped ground water, leaving 8 per cent to draw water from hand-pumped wells by 1991. One consequence of this bias towards piped supply from ground water sources is that diesel engines are needed to extract the ground water. Nowhere in the RWMPs are the implications of this decision, with respect to foreign exchange, skilled manpower, supply reliability and cost, made explicit or quantified. However, not far into the implementation period, "it was decided to build shallow wells wherever feasible for both economic and technical reasons" (FINNIDA, 1984, Annex 1,1). Hand-pumped wells now supply 70 per cent of the people served by the project since 1978.

Another important technology choice concerns the type of handpump to be used. The RWMPs (Finnwater, 1977a, 118) give first priority to a Tanzanian-made pump. No such pump existed at that time, and no efforts by the Project or FINNIDA to help to start local production appear to have been made. Instead, a Finnish company supplied the pumps right from phase I (Finnwater, 1980, 16). Its design was based on the company's pump for one-family use in Finland. Although continuously changed since 1978 it is difficult to maintain under rural conditions in Tanzania (see below). Experiments with a new hand-pump type are now being undertaken.[5] It may eventually replace the old type.

4.5. Service Level

The RWMPs proposed that schemes be designed to supply 30 litres per person per day through one water point per 200 people (Finnwater, 1977a, Ch. 6.1). During phase I of the implementation the service level was significantly lower: one well for each 300 people; one domestic point for each 500 people (Finnwater, 1980, 8 and 25). Later, the service level was improved for hand-pumped wells. Furthermore, piped supplies are now constructed according to the stipulated criteria (FINNIDA, 1984, 6 and Annex 4,2).

4.6. Priorities

The RWMPs do not specify any priority-weighting between regions. In reality it appears that donor funds have been allocated roughly

---

[5] The experiments with this new direct-action pump are done in cooperation with the Ministry and the World Bank.

according to population size (Ministry of Water and Energy/FINNIDA, 1981, 11). With respect to piped supplies, large schemes with low construction costs per capita, serving villages far from water sources, have been given high priority. The actual selection from the priority list is based on "their suitability for implementation by Finnish experts" (Finnwater, 1977a, 120; 1977b, 16). No criteria for selecting villages to be supplied with hand-pumped wells are given in the RWMPs (Finnwater, 1977a, 58). Neither are such criteria specified in later reports available. Thus it appears that the construction of wells, which is the main implementation activity, is not guided by any explicitly stated priority criteria.

4.7. Construction

In the RWMPs there are no proposals for rehabilitation, although the consultants found most existing schemes in poor condition (Finnwater, 1977c). The focus is on new schemes only. The main body of the RWMPs contains proposals for a crash construction programme (1977–1980) and an implementation programme (1981–1991). The aim of the former was to provide each village with one water point before 1981 in accordance with the new Tanzanian objective made in response to the villagization campaign.[6] The aim of the latter was to provide easy access to water to all villages by 1991. To accomplish this, the RWMPs envisage that the MAJI organizations at regional and district level should be expanded in order to double their implementing capacity (Finnwater, 1977a, Ch. 6, 7, and 8). There are no analyses of the feasibility of this proposal in terms of manpower and other required resources.

These proposals never materialized. The important chapter in the RWMPs is the last one in which the consultant puts forward "Proposals for Mtwara-Lindi Water Supply Project". Here it is stated that Finnwater should construct 500 hand-pumped wells as part of the crash programme in addition to the construction of 12 piped schemes during the period 1977–1979. It is this proposal (slightly modified) which FINNIDA finances.

Right from the start Finnwater became the turn-key contractor for the construction of these schemes. MAJI's role (and apparently that of FINNIDA as well) was reduced to one of being a passive observer.

---

[6] See Chapter 2. This aim was only pursued for a couple of years in the Finnish project.

Together with regional and district authorities they merely approved Finnwater plans. These plans were strictly production oriented and Finnwater had the ability to deliver the goods in time. Undoubtedly that contributed to the blissful atmosphere surrounding Finnwater until 1984. Few consultants involved in development projects get through an evaluation with high praise like the following: "The implementation of the project through the continued use of Finnwater has been the major reason for success to date, the efficient organization and professional approach to the task being of an extremely high order" (Ministry of Water and Energy/FINNIDA, 1981, 13). Short-term gains in productivity have resulted from these choices. But the long-term costs have been significant. Thus the choice of tractor excavators for well-digging mentioned earlier has several consequences.[7] Only locations accessible to tractors can be chosen as well-sites. Sometimes this conflicted with the easy access of women to the wells (FINNIDA, 1984, Annex 3, 20). Moreover, well depths were restricted to 4–5 metres due to the limited reach of the back-hoe with which the tractors were equipped. This is the basic reason why, in a normal year, about 750 (50 per cent) of the hand-pump wells will yield insufficient water during the dry season. The default figures will be higher in a dry year. "It is clear that a large number of wells are too shallow", despite the warning in the RWMPs that 25 per cent of all wells should be drilled up to 50 metres depth (FINNIDA, 1984, 13, 15; Annex 3; Table 4).[8]

4.8. Operation and Maintenance

This most crucial aspect of rural water supply development only receives scant attention in the RWMPs. There is no analysis of past experience; no specific proposals for improvements of the O&M system run by MAJI; no specific cost analyses of present and future requirements; and no proposals on cost-recovery to pay for O&M expenses. In the RWMP for Lindi region, for instance, fewer than three pages scattered in different volumes are devoted to the subject.

From a short-term perspective this neglect did not matter much, for

---

[7] Around 70 per cent of the 1750 wells completed by 1984 had been dug in this way (FINNIDA 1984, 14, 17). Tractor excavation ceased in 1985.

[8] There is another reason for the high rate of dry wells. In the RWMPs the extensive use of hand-pumped wells was not envisaged. Only limited hydrogeological surveys of shallow ground water were therefore made at that time. The later switch in technology from piped supplies to hand-pumped wells was not followed up by systematic hydrogeological surveys. The project therefore still has too little knowledge of the hydrogeology of the two regions, according to a Finnwater staff member interviewed in 1985.

Finnwater soon had to take responsibility for operation and maintenance of the many schemes they built with such speed and efficiency. Hand-pump wells were maintained by mobile teams with a Landrover. Each team comprised of an expatriate supervisor and four pump mechanics.

In recent years four visits have been made to most wells every year. At any one time 10 per cent of the hand-pumps were out of order and this is excellent by any standard (FINNIDA, 1984, 16). The corresponding figure for schemes maintained by the regional and district authorities is 75 per cent. That is, only 25 per cent of these installations supply water on a regular basis (Finnwater 1985b, 21). The problem of handing over Finnwater schemes to local authorities will therefore be enormous.

A maintenance system is now being slowly introduced. At the village level the shift started in mid-1983. Finnwater claims that it is fairly successful. The aim is to establish a three-tier maintenance system (village, district, region), with the main emphasis on the village level. This will eventually also involve MAJI. It will have to take over from Finnwater, but this change has not yet started. By focusing on establishing an O&M system for the FINNIDA funded schemes only, Finnwater is using a strict project approach in its attempt to solve the O&M problem.

Some formidable obstacles must be overcome to shift responsibilities to local institutions. The Finnish NIRA pump used so far is not suitable for village level repairs. It requires a large range of fairly complex tools for disassembly (FINNIDA, 1984, Annex 3, 23). A completely new pump-type is being tested and may later replace the NIRA pump on existing wells (Ch. 4.4). The regional and district MAJI authorities have presently no hand-pump maintenance expertise and little has so far been done to increase their capacity for this. Furthermore, "it is expected that the pressure on the budget will become excessive after completion and transfer of the current project works to local institutions" (FINNIDA, 1984, 54). In fact this situation *already* exists: the diesel available for operating schemes is only 10 per cent of demand in Lindi (where there are 100 schemes; FINNIDA, 1984, Annex 4, 5). The 1984 evaluation therefore rightly concludes that long-term FINNIDA commitment is needed to establish maintenance, although "it is not yet possible to define the scale and duration of this support". (FINNIDA, 1984, 43). A rather amazing statement after 13 years of continuous involvement in the rural water sector!

4.9. Participation

The RWMPs contain no proposal on participation. It is stated, however, that the "self-help principle" may slow down implementation; but also that it may "lower cost". Furthermore, "it should be considered whether ujamaa villages can be responsible for operation and maintenance" (Finnwater, 1977a, Ch. 7.5. and 15.1.). That is all. The RWMPs contain no analyses to back up these statements, nor any proposals on how villages could be made responsible for O&M in an operational way. Participation was clearly a non-issue to the water master planners.

Nevertheless, the word participation was often used in subsequent years when villagers were paid to dig trenches or wells, or when they were trained as operators and put on the government or Finnwater payroll. The 1981 evaluation raised a mild objection to this neglect (Ministry of Water and Energy/FINNIDA, 1981) and from mid-1983 a more systematic approach to participation has been introduced. With an energy typical of the project, villagers are now gradually being involved in planning, construction, operation and maintenance. It has and will slow down the implementation speed, and more hand-pumps will break down and not be repaired. But the 1984 evaluation has strongly supported the changes and they will continue in phase IV.

4.10. Coordination

The RWMPs contain no proposal or discussions of coordination between sectors or between the consultant and the Tanzanian authorities.

The 1981 evaluation noted that "little co-operation exists between different functional Ministries at the regional level"; and that "there is little coordination between regions in Master Plan Implementation" even in projects that cut across regional borders (Ministry of Water and Energy/FINNIDA, 1981, 9). Furthermore, there was no formal coordinating mechanism between Finnwater and the regional and district authorities. At the beginning of 1984 such links were being established through a Regional Steering Committee for water supply development which UNICEF[9] pushed strongly for (see also Ch. 4.3).

---

[9] UNICEF provided some of the funds for implementation through Finnwater (see Table 6).

4.11. Monitoring and Evaluation

There are no proposals or suggestions whatsoever on these issues in the RWMPs.

During the implementation phase quarterly progress reports and final reports for the three implementation phases have been made. However, it is not known "to what extent, if any, these reports have been used for project monitoring either by FINNIDA or MAJI" (FINNIDA, 1984, 26). This is, perhaps, a polite way of stating that the reports are inappropriate for monitoring purposes. They do not present any comparisons of plans with actual achievements. Neither do they provide any regular information about the functioning and utilization of completed schemes. The project management was running the construction activities without obtaining regular village level information about their effects.

There have been two external evaluations of the project. The first one was made in 1981 after three years' implementation and basically gave the project a blue stamp (Ministry of Water and Energy/FINNIDA, 1981). Three years later the second evaluation presented a critical analysis of the problems with the turn-key approach, as indicated above (FINNIDA, 1984). It helped to hasten the changes already initiated in 1983.

4.12. Key Observations

The RWMPs produced by Finnwater present an interesting mix of planning approaches. The plans are strictly functional in the sense described in Chapter 3.2. They contain no scrutiny of Tanzania's policies and goals for the rural water sector. These were simply accepted as a basis for planning. Second, the RWMPs are blueprint-oriented in the sense that they are prepared by planners *prior* to and independent of implementation activities (Ch. 3.3). The plans specify the future implementation rate until 1991, but apart from that they do not contain the detailed specifications of future activities implied by the blueprint approach. Third, the RWMPs are based on a set of disjointed analyses (Ch. 3.4). No systematic and comprehensive analyses of the proposals are made (except certain technical ones), and very few alternative solutions are considered. All non-technical issues are neglected in the RWMPs. The implementation plans which were prepared subsequent to the RWMPs have the same general features.

A non-participatory approach to planning and implementation was used from the start of the RWMP preparation until 1983, when some participatory elements were introduced. These aim at mobilizing the beneficiaries to take part in project activities (Ch. 3.5).

Finally, bypassing of local institutions right from the village level to the national level has been a key feature of the donor approach from 1972 until now. By the end of 1984 no systematic effort had been made to strengthen the institutions which will eventually take over the schemes completed by Finnwater.

Several lessons can be learned from this case study. The RWMPs appear to have had only limited relevance during subsequent implementation. Thus, the RWMPs proposed a technology (piped schemes) that was subsequently only used to a limited extent. They did not provide priority criteria for the hand-pump well construction, which was by far the most important construction activity. They warned against building wells too shallow but the warning was ignored. They assumed that construction, operation and maintenance should be carried out by existing MAJI organizations and that their capacity could be rapidly increased. Yet this was never attempted. All project activities were taken over by consultants. The RWMPs contain very few or no proposals on operation and maintenance, participation, coordination, monitoring and evaluation. They have, however, been useful in giving a general picture of water resources according to observers (but see Ch. 10.6). All in all the obsolescence of the RWMPs is perhaps best illustrated by the lack of specific references to them in subsequent project reports.

The case also illustrates how a strong emphasis on production of new schemes to reach ambitious targets has displaced long-term development considerations. The production bias has, furthermore, had major consequences for the organizational set-up of the project; for the choice of construction methods; and for village participation and institutional development. Short term results have replaced considerations about the long-term sustainability of activities.

The relative isolation of the project from its initiation to around 1984 is noteworthy. In this period the project management had little contact with similar projects in other regions. It also suffered from a lack of any regular village level monitoring of project activity outcomes. It did not, and could not, learn from its own mistakes and those of others.

It is also instructive to note how the Finnish approach to planning and implementation has remained basically unchanged from 1972 to 1983. No significant changes took place during this period. When they occurred, they were introduced by a new project management and an external evaluation. No internal learning process resulting in a gradual project development took place for 10 years.

Finally, the very limited role of the Ministry during the entire project period is remarkable. It has neither taken a strong interest in a regular monitoring of the project activities, nor did it play an active role in changing the project approach in 1983 and 1984. It appears to have been satisfied with the fast rate at which wells were constructed.

5. The Turn-Key "Approach" with a Salesman's Touch: The Dutch in Morogoro Region

Dutch assistance to the rural water sector, like the Finnish, passed its 10th anniversary some years ago. Certain key aspects of their approaches are similar. The Dutch have also put considerable emphasis on the production of new schemes, mainly shallow wells.[1] They have kept to the same consultant, DHV, throughout the years, first in Shinyanga region from 1971 to 1977, and then in Morogoro region from 1978 and onwards. Furthermore, in both regions – but especially in Morogoro – DHV has run a project implementation unit independently of the existing MAJI structure. However, the Morogoro project has a number of unique characteristics which are interesting from a planning and implementation point of view:

– a Water Master Plan for the entire region has not been prepared; instead, medium and long-term plans for drinking water supplies for specific high-need areas have been worked out;

– the assistance has been divided into numerous separate projects implemented by the consultant;

– the consultant, supported by the donor, has aggressively promoted the shallow wells technology in the whole of Tanzania;

– since 1978 repeated attempts at handing over completed wells to local authorities have failed although this transfer was stipulated in government-to-government agreements.

Both piped water supplies and shallow wells were built with Dutch assistance. In the following analysis the main focus will be on planning

---

[1] The terms "shallow well" and "hand-pumped well" are used interchangably in Tanzania. The Dutch use the former term which has therefore been adopted in this chapter.

and implementation of shallow wells activities, as they can best illustrate some of the key problems in the chosen donor/consultant approach.

5.1. Description of the Project, 1978–1985

Well construction in Morogoro region is in many ways a continuation of the Dutch-financed Shinyanga project. Therefore a brief description of the latter is needed to put the former into perspective.

Dutch involvement in the Tanzanian rural water sector started in Shinyanga region in 1971 with the preparation of a RWMP. In this plan it was proposed, among other things, that 2000 hand-pumped wells should be constructed. Subsequently the Dutch, through DHV, financed the construction of 714 hand-pumped wells from 1974 to 1978. During this period they also established a four-tier well maintenance system. Throughout the project period these activities were run by DHV through a fairly independent project management unit. Local institutions from regional to village level were not much involved (Ausi, 1979). As far as village participation was concerned this was contrary to the initial "very high" expectations of involving users. But "due to production pressure and problems in organizing free labour, village participation was reduced to "partly paid labour" during construction" (Hordijk *et al.*, 1982, 35). Employment of counterparts for the Dutch professional staff was also stipulated in the Dutch-Tanzanian agreement, but only one or two were attached to the project and even that on a irregular basis. Tanzanian professionals were simply not available or they were sent to other regions (see Ch. 2.3).

By 1978 the Dutch assistance to water supplies in Shinyanga region stopped and DHV moved to Morogoro. The RWE, Shinyanga, took over the equipment, vehicles and Tanzanian staff trained by DHV. He was then supposed to continue constructing wells and maintaining the 714 already completed. But DGIS and DHV had made no systematic efforts to prepare the RWE organization for this responsibility, and the Tanzanian authorities only provided the RWE with limited funds. Consequently, the construction rate fell to one-fourth of DHV rate and only about half of the already completed hand-pumped wells were in working order by the end of 1981 (Anderson, 1982, 13, 30). As one Dutch consultant, who was involved in both Shinyanga and Morogoro, remarked in an interview: "The RWE did not have the resources to take over the wells. It was crazy to try".

When DHV moved to Morogoro in 1978, this clear evidence of the non-sustainability of the Shinyanga activities had not yet become available.[2] In fact, both DHV and DGIS assumed that the approach used in Shinyanga was basically sound. So, apparently, did the Tanzanian authorities. As in Shinyanga, DHV was to put the main emphasis on the *production* of wells. However, some new and important changes were made when DHV started up in Morogoro.

First, the Dutch decided – despite Tanzanian objections – not to prepare a comprehensive RWMP. Based on a short initial survey of ground water resources, DHV wanted to start the construction of hand-pumped wells right away, while preparing long- and medium-term plans for drinking water supply in high-need areas only (see later). Second, both DGIS and DHV saw the hand-pump technology as *the* answer to Tanzania's rural water supply problems. They wanted to promote it, not only in Morogoro, but nation-wide. Thus with Dutch assistance Morogoro became the headquarters and DHV the contractor for (i) the preparation of a National Shallow Wells Programme (see below); (ii) the training of MAJI and donor staff from all regions in construction and maintenance of shallow wells; and (iii) the assembly and distribution of hand-pumps and drilling equipment, imported from Holland, to all regions.[3] Third, both DGIS and DHV wanted to increase the efficiency of the DHV operations. Therefore, DHV worked completely independently of the RWE in Morogoro.[4] The fourth new element in the Morogoro approach was that the RWE should maintain the wells completed by DHV. These wells would be handed over to the RWE every three months to avoid the abrupt transfer of wells that took place in Shinyanga when DHV left.

The Dutch assistance in Morogoro has been split up in numerous separate projects, of which the well construction project is the largest. The projects are listed in Table 7 together with their costs. A total of 810 hand-pumped and 22 piped supplies have been constructed by DHV. If functioning and utilized as designed, they supply approximately 305,000 people. Until recently 12 to 16 expatriates and 60 to

---

[2] In 1982 DGIS committed 2.6 mill. Guilder to the rehabilitation of wells in Shinyanga. By 1985 it intended to place two expatriate advisors within the RWE office in Shinyanga to help in the rehabilitation and re-establishment of a maintenance system.
[3] DHV also hoped that this promotion of shallow wells in the public sector would eventually open the opportunity for promoting similar activities in the private sector. Interviews with DHV staff, 1985.
[4] Symbolized by the complete physical separation of the MAJI and the DHV workshop and yard facilities.

Table 7. *Dutch Funding and Output of the Morogoro Projects, 1978–1984*

	Implemented since	Total[a] commitment	People served (×1000)
Project Preparations			
Morogoro Domestic Water Supply Plan	1978	4,9	
Morogoro Gravity Project	1979	1.0	
National Shallow Wells Programme	1980	0,8	
Management and improvement of population participation in rural water programmes	1983	0.5	
Rural Water Supply Survey in Southern Morogoro	1980	1.8	
		9.0	
Implementation			
Morogoro Wells Construction Project (incl. training)	1978	18.5	202
Morogoro Piped Water Supplies	1979	15.4	
		33.9	103

a Mill. Dutch Guilders as of Jan. 1, 1984.
Sources: WHO (1985) and interviews with project staff.

130 Tanzanians have staffed the projects, but staffing is now being drastically reduced.

This reduction in DHV staff was caused by the criticism of DHV, DGIS, and the Tanzanian authorities raised by an independent Tanzanian-Dutch evaluation team in early 1982 (Hordijk *et al.*, 1982). The evaluation has led to a virtual standstill in the construction of new schemes. It has also motivated a drastic rethinking of the approach to the Dutch assisted rural water supplies. The period from July 1982 to the end of 1984 has been a crucial transition period for the project (DHV, 1982). But by the end of 1984 a Tanzanian-Dutch study team had worked out a new approach to rural water supply development in Morogoro region.

According to this proposal it is the intention to place the responsibility for the Dutch-funded activities with the Tanzanian authorities from village to national level in the full knowledge that their capacity to implement is low. The responsibility for operating and maintaining schemes should be placed at village level, and DHV should have a much less prominent role. Expatriates should have advisory – not executive – responsibilities within the RWE's and DWE's offices. In short, the ball should be put in the Tanzanian court – with the villages and with the recipient authorities (Anon, 1984).

The initiative in proposing these drastic changes has come from DGIS. Both DHV and the Tanzanian authorities – especially the regional politicians – have strongly resisted them. Obviously the new approach will result in drastic reductions in construction rates. DGIS hopes it will also result in more sustainable benefits of the Dutch assistance. However, by the end of 1985 the future of the Dutch projects in Morogoro was still uncertain.

Given the basic problems experienced, first in Shinyanga and then in Morogoro, it is understandable that drastic changes in approach *had* to be made. That it should take more than 10 years is less understandable. It indicates some basic flaws in the donor-consultant-recipient framework for aid. Some of them will be analysed below.

5.2. Planning

The Morogoro Domestic Water Supply Plan (MDWSP) was to provide the guidelines for future activities in the northern part of the region.[5] DHV prepared this and all subsequent plans in the turn-key mode – independently of the recipient organizations. The plan is based on the premise that the 1991 target should and could be reached. It contains substantial information on water resources and on the requirements for manpower and construction facilities. It is, however, "not based on any consideration of the implementing capacity" of the village, the district or the region itself. There is "no integration with the local and regional decision-making structures". The projects as proposed are therefore "additional to what is done locally and can be seen as actual gifts from the outside agency." (Hordijk *et al.*, 1982, 14).

It was originally intended that the MDSWP was to form the basis for the preparation of an implementation programme:

...due to the fact that other projects [for instance the wells construction project; see also Table 6] were started before this programme was formu-

---

[5] The Ministry has repeatedly (but in vain) requested DGIS to support the preparation of a water master plan for the entire region.

lated, the whole planning process has become a series of "ad hoc" projects. It appeared that the implementation of the projects "lived their own lives" and were extended one or more times because they were difficult to stop ... a mutually agreed and integrated strategy for development of rural water supplies in Morogoro region has not been formulated yet (Hordijk *et al.* 1982, 24).

Another interesting exercise in policy-making and planning has been carried out within the National Shallow Wells Programme (NSWP) funded by the Netherlands from 1980. It was to be a study of the organizational, economic and participatory aspects of promoting this technology on a nation-wide scale. A final report has never been made, for the mixture of shallow wells and participation soon proved to be a volatile one, involving hidden agendas and conflicting interests between the agencies involved: DHV, DGIS, the Ministry and IRC/PMO[6], who were consultants on the community participation aspect of the NSWP.

DHV as a company had a legitimate business interest in promoting the shallow well concept and the company's own role in a nation-wide programme. This it did very aggressively. As the Project Director in charge of the NSWP explained to a gathering of donor and Tanzanian representatives involved in rural water supplies:

> Until 1991 all efforts should be directed towards implementation of ... shallow wells only. If 75 per cent of the rural population can be supplied ... then the remaining 25 per cent ... will voluntarily move to those areas where shallow wells have already been constructed .... Until the required number of wells has been constructed execution must be carried out on a contract basis ... directly funded by donors; not as part of the RWE organization; not under government regulations .... [One organization should] ... test systems; materials and equipment; provide guidelines for all regions; procure and distribute all materials and equipment .... The Morogoro Wells Construction Project can undertake this for the time being (Bonnier, 1980, 20–21).

Next there was a donor, DGIS, that appeared to support the idea of shallow wells and the consultant's role in it as *the* solution to the rural water problems. The evaluation report summarized it as follows (Hordijk *et al.*, 1982, 51):

---

[6] IRC is funded by WHO.

The NSWP so far had caused confusion with the Tanzanian authorities as to its status and objectives. Is the Netherlands promoting the specific shallow wells, which are developed by DHV and seen as a commercial product of DHV? Or is it a real attempt to assist in establishing a balanced and appropriate rural development policy?

Then there was the Ministry of Water and Energy. The Minister, Al-Noor Kassum, had this to say (Hordijk *et al.*, 1982, 60):

> Of course we understand that DHV as a commercial firm tries to promote shallow wells, but why cannot the embassy present a more independent view?

But the Tanzanian authorities also had other problems with the shallow well technology. Although shallow wells were accepted wholeheartedly in public (Ministry of Water, Energy and Minerals, 1980a), and this acceptance was reiterated to the donors, there has been considerable resistance to the technology both within and outside the Ministry. To many, shallow wells are a temporary solution necessitated by resource constraints, the pressure to reach the 1991 goal, and the preference for this technology among many donors. The real solution to rural water supply problems, according to this view, is piped supplies. These are regarded as technologically "superior" and can provide a higher level of services because they make house connections possible (which a shallow well does not). Therefore many Tanzanian politicians, policy-makers and civil servants have an ambiguous attitude towards the shallow well technology. Thus, during discussions about the NSWP in early 1982 the Minister of Water and his Principal Secretary stated that although the NSWP study could continue "no decision as to a programme had been taken" (Hordijk, *et al.*, 1982, 50). By 1984 a decision to go ahead with a national shallow wells programme had still not been made (Ministry of Water and Energy, 1984).

Finally there was the IRC in the Hague, which DGIS invited to assist in the study of the participatory aspects of the NSWP. The ambitions of the IRC went beyond shallow wells. It prepared a project "profile" for the development of a community participation component. This profile does not indicate what the relation to the NSWP would be (IRC, 1980). In fact IRC soon started to operate as an independent project to "provide support to Tanzanian departments, operating agencies, and other relevant institutions to strengthen the element of community participation in water supply and sanitation projects" (IRC, 1981, 3). Moreover, from mid-1982, field testing of a community participation component started in cooperation with the Community Development Department (CD) in the Prime Minister's Office in Dodoma and funded by DGIS. In this way the CD and PMO soon found that the Dutch sponsored project had placed them in a central position in the rural water sector. Since then there has been some tension between MAJI and PMO about the responsibility for integrating participation in the rural water sector activities. This tug of war appears to continue.

In the end, the NSWP never produced a final report. DHV prepared four chapters on organizational and technical aspects of a national shallow wells programme. The chapter on maintenance was to be worked out by DHV and IRC in cooperation. This never happened. Instead, PMO/IRC (1984) wrote their own final report. And by 1984 the Ministry itself produced a new assessment of a nationwide Shallow Wells Programme. It contained no mention of the Dutch NSWP whatsoever (Ministry of Water and Energy, 1984). The donor/consultant attempt to prepare a national policy on shallow wells for Tanzania had failed, although it did help to put this technology on the national agenda for the rural water sector.

The Water Supply Survey, Southern Morogoro Region constitutes the third Dutch attempt at medium- and long-term planning. It was initiated in 1982 and finalized in October 1983. It contains, no doubt, valuable information on water resources, etc., but the criticism raised by the evaluation team early in 1982 and the increasing difficulties in maintaining already existing hand-pumped wells (only 57 per cent were operating successfully in early 1982 according to Hordijk et al. (1982, 16)), had no discernible influence on the planners. In the Water Supply Survey they only touched the operation and maintenance problem in a few sentences while proposing to construct an additional 550 wells (DHV, 1983, Ch. 8).

This donor and recipient bias towards production targets combined with a narrow project approach can undoubtedly be traced both to the Terms of Reference, the consultants' contract with DGIS, and the sympathy of the Tanzanian authorities for construction activities. The bias has had pervasive influences on what was implemented in Morogoro region and how this was done as will be described below.

5.3. Institutional Framework

Until the end of 1984 there was considerable confusion with respect to the institutional framework for the Dutch assistance. This confusion exists to varying degrees in all five case study regions reported here, but perhaps particularly in Morogoro. Some of the main causes merit comments. Most of them are not specific to DHV and DGIS.

In the case of shallow well activities the institutional framework was partly determined by the "Agreement of Technical Cooperation" of 1965 between the Netherlands and Tanzania. It requires that an administrative agreement between the two governments is made, outlining the aims, responsibilities and contributions, etc. of the two parties. Such an agreement was never signed for the shallow well activities (Hordijk *et al.*, 1982, 42 and 57). Furthermore, the technical cooperation agreement also requires a "delegation contract" between DGIS and an executive agency – in this case DHV. The Tanzanian authorities have had no formal part in this contract which is signed by the donor and the consultant only. The contract delegates "all activities within Dutch assistance to shallow wells activities to DHV, hereby designating the consultant to be both advisor, trainer, supervisor and contractor at the same time". However, contracts between DHV and DGIS were unclear and incomplete. Although construction targets were specifically stated, "thorough project descriptions, schedules of operation, etc., were . . . non-existent." (Hordijk, *et al.*, 1982, 4,7)

Such a plethora of ill-defined roles obviously enhances the position and influence of the consultant, and the way DGIS carried out its role as a client did little to counterbalance this. DGIS never developed a clear policy or explicit priorities for its assistance to rural water supplies in Tanzania.[7] And the DGIS-mission staff in Dar es Salaam was just "very eager to get things done". It initiated many activities, became operationally involved in the various Morogoro Projects and developed close contacts with the consultants. As already discussed, this intermesh of consultant and DGIS interest caused some resentment on the Tanzanian side. In their eyes "DHV was the last part of the Dutch Development Assistance arm in the region". This suspicion was strengthened in the region, by, for instance, the refusal of DHV to let RWE-staff read the contract. The close cooperation between DHV and the DGIS-mission in Dar es Salaam also caused serious tensions

---

[7] One reason for this could be that a thorough reorganization of DGIS headquarters led to increasing workloads and rapid turnover of the staff dealing with the rural water projects.

with DGIS in the Hague. Headquarters was "in many cases just asked to endorse what had been decided or had happened already" (Hordijk et al., 1982, 55–60).

This institutional confusion on the donor side was exacerbated by problems on the Tanzanian side. Water activities are divided into small "regional projects" for which the regional authorities and the RWE are responsible, and large "national projects" falling under the Ministry (see Ch. 2). The amount of Dutch funds channelled into construction of wells qualified the project as national and therefore it was the Ministry which became the counterpart organization for the Dutch project. However, the Ministry never took an active interest in the shaping and monitoring of the consultants' and the donor's activities. On the other hand, once shallow wells are constructed, they fall under the RWE, who is in overall charge of maintenance. But the regional authorities were largely excluded from influence on the project. They did not receive sufficient local or donor resources to carry out the necessary maintenance activities. The RWE was constantly requesting vehicles, spare parts and fuel, etc., without result.

To sum up: DGIS signs an agreement with the Government of Tanzania which in turn designates the Ministry of Water to become the responsible recipient organization for the shallow wells activities. DGIS also signs a contract with a consultant to which the responsibilities for all activities are delegated. DHV, in effect, becomes the jack-of-all-trades in the project. The RWE, the regional authorities and the villages are not involved *until* the completed wells are to be handed over to them. Even a direct formal channel through which the regional authorities and the consultants could discuss was only established in early 1984. In the words of Kauzeni and Konter (1981, 35):

> The Tanzanian authorities had no say, whatsover, in the set-up and execution of the programme.

In 1984 a study team recommended that this bypass approach should be completely changed. It proposed, as described in Chapter 5.1., that Dutch technical assistance and funds should be channelled to the regional and district water authorities and that Dutch assistance should be adjusted to fit the local capacity to construct and maintain (see also Ch. 5.8 and 5.9). The execution of this proposal had not yet started by the end of 1985. DHV naturally fought to keep a role – even a reduced one in Morogoro. There was also some strong Tanzanian opposition (see Ch. 5.12).

5.4. Scheme Technology

The choice of technology is clearly linked to the choice of rural water policy in general and to operation and maintenance policy in particular. DHV has, over the years, put considerable effort into developing techniques that could reduce construction and maintenance costs of hand pumps and wells and increase construction speed and thereby increase the possibility of reaching the 1991 goal. A main consequence of these efforts has been that the techniques developed for wells and hand-pumps have required an increasing input of specialized skills in survey, construction and maintenance. "The technology has been moved away from the village" (Hordijk *et al.*, 1982, 32). This has obvious implications for the organization of an O&M system (see Ch. 5.8).

Furthermore, the new techniques (and the economic crisis in Tanzania) required that an increasing share of hand-pump and well parts was imported from the Netherlands. Local manufacture of hand-pumps, which was originally intended, never materialized.[8] It does not appear to have been pushed very hard by any of the main parties involved (Ministry, donor, consultant). Thus, some Dutch volunteers, who were working on the project, and who attempted to develop a pump from local materials that could be maintained by the village, concluded that they did not "get much cooperation in order to experiment with new kinds of pumps. Meanwhile the expensive pumps were imported from the Netherlands, which were not 'maintenance free' at all as was proved later on" (quoted in Kauzeni and Konter, 1981, 33).

From the earlier emphasis on hand-dug wells lined with concrete rings and equipped with a hand-pump partly manufactured from locally available materials (the Shinyanga pump), the wells constructed in Morogoro were frequently drilled, lined with plastic pipe, and equipped with a hand-pump made exclusively of imported parts (the Kangaroo and SWN pumps). Unfortunately, this emphasis on technical solutions has not contributed much to an improvement in supply reliability of completed shallow wells as discussed below.

It belongs to the full picture of water supply technology development in Tanżania that DHV has succeeded in introducing its shallow well technology and hand-pump models in other regions. Thus, DHV has trained many shallow well construction teams from other parts of the country. It has also become the main supplier of hand-pumps and equipment to other donors in the water sector except Finland (see Ch. 4).

---

[8] Instead, hand-pump parts are imported and then assembled in Morogoro.

5.5 Service Level
In the Morogoro project a service level of a maximum of 250 people per well and a maximum walking distance of 400 metres to the well was aimed at. This level was not achieved in practice. According to a survey by Kauzeni and Konter (1981, 25 and 29), only 39 per cent of the villages provided with wells meet the first part of the criteria. No information on the distance criterion is available. The gap between actual and planned service level may arise because suitable sites for wells in a village are insufficient. In this case, as Kauzeni and Konter (1981, 30) concluded, "an adequate drinking water supply cannot be provided for by shallow wells alone". The gap may also arise because the consultant/donor/Ministry aimed at maximizing either the number of villages "covered" (with one or a few wells) or the number of wells constructed (regardless of their location in relation to house settlements). Whichever the case, when the distance to a well or the queue becomes too long, women will go to streams, waterholes or other sources of water. Only recently is this problem being addressed, but in other regions (see Ch. 2.1 and 8).

5.6. Priorities
Need criteria (distance; water quality, quantity and reliability; village population size) together with consideration of the logistics of construction have been used to identify priority areas in the plans for both the northern and southern part of the Morogoro region (DHV, 1983, 35). To what extent actual construction has followed these criteria is not documented.

It is, however, clear that the regional authorities have had no formal say in the selection of villages, although DHV's yearly construction programme has, according to DHV staff, been discussed with the RWE and the DWEs on an informal basis.

5.7. Construction
DHV and the RWE each ran their own construction programmes for shallow wells in Morogoro. The two programmes were executed fairly independently of each other, although DHV has trained some RWE staff. Only the DHV construction programme is discussed here.

The consultant set up a "very efficient organization . . . The targets as stated by the contract were reached in a reasonable time" (Hordijk, et al., 1982, 4). Emphasis on production was so strong that in line with present policy, the entire senior staff consists of expatriate personnel. Counterpart staff will not be appointed or seconded ... to senior staff, to be trained and prepared for a gradual take-over of their duties, until all survey, construction, production and supply activities are operating according to plan and at required output level. (DHV, 1979)

Due to the many expatriates in the project, construction costs have been high. The Tsh 20–25,000 per hand-pumped well normally quoted by the consultants (Bonnier, 1980, 18–20) exclude overheads. When these are added, the real financial cost varies between Tsh 60 and 70,000 per well depending on the method of calculation. Almost 70 per cent of this cost is paid in foreign currency. If the same wells were to be constructed by the Tanzanian authorities the cost would be around Tsh 30,000 per well. As concluded by the evaluation team:

The level of technology and management complexity ... of shallow wells ... is not that advanced to warrant the involvement of so many expatriate experts. (Hordijk *et al.*, 1982, 24; 66–73)

With the blessing of the Ministry and the donor, DHV continued with the bypassing approach from 1978 to 1982. During this period the focus on production targets was combined with efforts to develop a high and uniform standard of wells, hand-pumps, and drainage facilities around the wells. The end product was of a high technical quality and was, furthermore, very well documented. But the control of all project activities remained firmly in the hands of DHV. One junior level expatriate working on the assembly of pumps in 1981 wrote:

I had no Tanzanian counterpart and no serious efforts were undertaken in order to provide one. When I took some initiative in order to pave the way for a handing over, I got the hint that I should stop this. (Quoted from Kauzeni and Konter, 1981, 42).

5.8. Operation and Maintenance
The "handing-over" of completed wells to the Tanzanian authorities became an ever-growing problem as construction went ahead as planned. This was the Shinyangya problem over again, according to the DHV project manager who was in charge of well construction in both Shinyangya and Morogoro. He noted that in the terms of reference for the Shinyangya project it was written that:

> ... the project should maintain the wells until the end of the project period. Full stop. Who would take care of the wells was not specified. (Van der Laak, 1980, 31)

In the Tanzanian-Dutch documents guiding the Morogoro project, a mechanism for the transfer of wells to the Tanzanian authorities *had* been better specified: every three months the wells completed by DHV were to be handed over to the RWE. This never happened. By July 1981, for example, after three years' production, 200 completed wells had not yet been transferred. This was a frustrating and potentially dangerous situation. Over a long period the Dutch pushed hard to change it and at a meeting in May between DHV, the RWE and the regional authorities it was agreed in writing that all completed wells *should* be transferred to the RWE.

> At the fixed time and location no Tanzanian representatives turned up, however, and at ... the end of July 1981 ... not a single well had yet been transferred to the RWE. (DHV, 1981, 42)

The RWE simply did not have the capacity to take over the DHV wells, neither had he nor other Tanzanian authorities been much involved in the programme and its execution. The RWE did not even have sufficient funds to maintain the schemes constructed *prior* to the arrival of DHV.[9] Thus, there was no coordination between the RWE's own capital and recurrent budget – let alone any coordination between the construction budget of DHV and the recurrent budget of the RWE. A promised contribution to the RWE's recurrent budget from DGIS of 300,000 Dutch Guilder would not solve the RWE's problem in the long run.

Although some staff from the regions and districts were trained by DHV, "No definite maintenance system ... was ... set up" (Van der Laak, 1980, 32). This was still the situation in early 1985, but a number of initiatives to rectify the situation were under way – strongly supported by DGIS (see Anon, 1984). However, the actual costs of maintenance also appeared to be increasing. The early DHV estimates of 1 per cent of construction costs (Bonnier, 1980, 18) had been adjusted upwards to 7 per cent three years later (DHV, 1983, 157). If depreciation is included the figure is 20 per cent (DHV, 1984). All

---

[9] In 1982/83 the RWE would need a doubling of current funds to do this, according to a Dutch interviewee.

these costs relate to the construction costs excluding overheads (see above).

Increases in the estimates of maintenance costs also indicate that the pumps installed by DHV needed regular repairs. Only 57 per cent of the wells investigated by the evaluation team in 1982 provided water of an acceptable quality throughout the year (Hordijk *et al.*, 1982, 16). And during the rehabilitation of wells by DHV in 1983–84 it was found that 85 per cent of the wells had one or more major problems although only 5 per cent were dry. The widely used Kangaroo pump will develop a major problem every five years (DHV, 1984, 5). Perhaps the repeated claim by the consultant (Bonnier, 1980) that it was possible to develop a "maintenance-free" pump, and that the consultants were doing it,

> has given some Tanzanian authorities an argument, why there was no need to consider pump maintenance as from the start (Hordijk *et al.*, 1982, 33)

5.9. Participation

DHV (1980, 3) itself described the prevailing situation very well:

> There is at best nominal participation (e.g. in the form of paid self-help, a contradiction in terms), a brief consultation as to the preferred site of wells and a single meeting during which the importance of clean water and the need for maintenance are explained...An unpaid pump attendant is trained. Yet, once the well has been constructed, the villagers are generally left to their own resources.

But DHW was not contractually obliged to put efforts into participation and the donor and recipient did not press the point. When DHV was criticized for the lack of participations they argued that "the authorities don't believe in the feasibility of village participation" (Hordijk *et al.*, 1982, 35). In an interview in 1985 the project manager for DHV added:

> The minute you start with participation your targets fall to pieces...Our targets were pretty tough; both donor and recipient insisted on them...We were contractors.

The PMO/IRC proposals on participation, originally intended as a part of the National Shallow Wells Programme, did not become integrated in the operational aspects of the Morogoro Wells Programme (see Ch. 5.2), and did not contribute to a change in the non-participatory approach until DGIS and the Ministry in 1984 set up a study team to prepare a complete change in approach (Anon, 1984). It is this new participatory approach which will perhaps be implemented in the future. According to the new proposal, the beneficiaries will participate in the planning, construction, operation and maintenance of water supplies. They will also be required to contribute in cash and kind to cover O & M expenses. Thus, the proposal implies a break with the Tanzanian policy of water as a "free" public service. By the end of 1985 the new approach was still being discussed (see below).

5.10. Coordination

It should be clear from the discussion above, that DHV operated as an independent contractor without much coordination with the Tanzanian authorities. Increasingly this has contributed to the feeling that "the Dutch parachuted the project into the region" (Hordijk et al., 1982, 54). This problem is now being addressed. A formal steering committee will be set up to direct Dutch assistance to water development in Morogoro. Furthermore, the executive role of Dutch expatriates will be abolished. How this will work out in practice remains to be seen.

5.11. Monitoring and Evaluation

Throughout the period from 1978 DHV has regularly produced progress reports. For specific comparisons of what was planned and what was achieved they are not particularly useful, perhaps because the project has never developed explicit project descriptions and plans of operation except with respect to production targets (Hordijk et al., 1982). On the other hand the consultants' reports were quite frank about the numerous problems encountered as a result of the production-oriented approach. The many DHV quotations used in *this* report clearly testify to this. Yet the consultants' warnings remained cries in the wilderness. For a long time neither donor nor recipient appeared to pay much attention. They did not act on the lessons and errors so clearly described. Monitoring had not much impact on project development either. Project activities were set in motion and proved hard to stop. Perhaps nobody really wanted to stop them (see below).

Changes first began when independent evaluators analysed the

Dutch assistance. Kauzeni and Konter (1981) and Hordijk *et al.* (1982) contributed significantly to this. Until then the Dutch approach had remained basically unchanged for more than 10 years.

5.12. Key Observations

In Morogoro region the Dutch have assisted in the preparation of long-term water supply plans for high-need areas only. No complete RWMP has therefore been made. The various partial RWMPs and subsequent implementation plans do, however, share the same key features as those RWMPs and implementation plans prepared by Finnwater for Mtwara-Lindi regions (see Ch. 4.12). Thus the Morogoro plans are functional and blueprint oriented and based on partial analyses only. They are biased towards water resource and technical issues. Furthermore the planning and implementation approaches used by DHV are strictly non-participatory. The Dutch assistance has largely been channelled through the consultant, DHV, thereby bypassing recipient organizations from village to national level with respect both to planning and to implementation activities.

This case study highlights the sustainability issue in donor-assisted rural water supply projects in Tanzania very well. It illustrates how four factors in particular have shaped the planning and implementation approaches of Dutch assistance. These are (i) the Tanzanian water-for-all-by-1991 policy; (ii) the Tanzanian policy of water as a free public service; (iii) the donor interest in assisting rural development activities that show immediate results and thus justify increased expenditure; and (iv) the consultant's interest in producing tangible results and expanding business.

A strong emphasis on *production of new schemes* has resulted from the coalescence of these factors. It satisfies the immediate interests of all three main actors. This production orientation has displaced other objectives that the donors, the recipient, or the consultants may have written into common agreements and plans. Even donor conditions written into government-to-government agreements (e.g. handing over of completed wells) were neglected once the technical assistance staff was in the field, the money committed, and the ball rolling. Nobody knew this better than the consultant. Thus DHV states (1980, 3):

> It would appear that success or failure of especially foreign-sponsored projects is measured solely in terms of the number of wells constructed and similar quantitative targets . . . . It is a sad thought that despite a much

greater effort from both donors and Tanzanian authorities during the past few years, the more fundamental issues of effectivity and impact still have not been raised other than in general statements at conferences etc.

When short-term production results are high on the common agenda, it follows that implementation through a contractor that bypasses local institutions becomes acceptable. This makes the donor and the consultant less dependent on the local institutions. Their capacity to implement is limited – a fact well known and regretted by ambitious Tanzanian decision-makers. However, the case study also shows that considerable costs can be involved in bypassing. It may result in short-term gains in production efficiency, but serious problems arise when – sooner or later – activities *must* be integrated into local institutions. Learning to operate within the local context right from the start may be more efficient in the long run than changing horses in midstream. Once the initiative and responsibility for planning and implementation has been taken over by the donor/consultant (for whatever benevolent or not so benevolent reason) it appears extremely difficult to get it back where it belongs: with the recipient organizations and the beneficiaries.

Furthermore, objectives about beneficiary participation in rural water supply activities have been displaced in the Morogoro projects. If water supplies are given as a free service there is only a limited scope for participation. Involving beneficiaries in project activities will also slow down the implementation rate. Therefore DHV practiced only "nominal participation", while it vigorously attempted to develop a maintenance free hand-pump that could make the project independent of any village involvement in operation and maintenance.

Finally, it is noteworthy that the complete changes in the approach proposed by the study team in 1984 originated outside the project – first from the evaluation team and later from DGIS. However, the DGIS change in attitude does not originate from the evaluation only. It is perhaps also related to a change of Dutch government and its generally more reserved attitude to aid to Tanzania.

6. The Pre-Preplanning "Approach": The Swedes in Lake Regions[1]

Sweden has assisted Tanzania's rural water sector development on a continuous basis since 1965. This has given her a major influence on sector development through a substantial financial country-wide support for new schemes and sectoral infrastructure; through Swedish manpower and planning assistance to the Ministry and to various RWEs; and through the funding of training and education of a large number of Tanzanian water technicians and engineers. By 1983 Swedish aid to the rural water sector amounted to 480 mill. Swedish kronor (SIDA, 1983b). It reflects Sweden's position as the biggest donor in the sector.

In this case study only the Swedish assistance to the Lake Regions (Kagera, Mara, Mwanza) will be analysed. For it is here that Swedish support has been based on extensive donor funded planning – starting with the preparation of Regional Water Master Plans in 1975. To avoid repetition of material from all three regions, the analyses focus mostly on wells development in Mwanza region.[2]

Swedish aid to the Lake Regions is interesting because:

- implementation (starting in 1984) has been preceded by almost 10 years of sequential and major planning exercises (preparation of RWMPs followed by implementation planning);

- the plans reflect the drastic changes in Swedish policies for aid to rural water sector activities made during this period;

- the present implementation approach is the most ambitious among

---

[1] The label "pre-preplanning approach" does not imply that it was deliberately chosen by the donor. It reflects the fact that substantial resources were spent on planning from 1975 to 1984 without any implementation. A change in the donor water sector policies has been a major reason for this.
[2] This also allows comparisons with the World Bank assistance to shallow wells activities in Mwanza (see Ch. 7).

donor funded projects; it is based on participation and integration of health, sanitation and water activities;[3]

- not only the construction of new schemes, but also strengthening the capacity of local institutions to plan and implement are now major objectives;

- the PMO – not the water authorities – is gradually becoming the coordinating recipient agency at field level.

The case study illustrates the labour pains involved when a donor makes drastic changes in its entire approach to aid to rural water development.

6.1. Description of the project, 1975–1985

For approximately 10 years from 1965 onwards Sweden financed around 80 per cent of all new rural water schemes on mainland Tanzania. No support for operation and maintenance was given. A substantial part of the schemes was diesel-pumped supplies. The Swedish funding coincided with the take-over from 1965 by central government of all construction costs of rural water schemes. Hitherto this cost had been covered by local governments. On a country-wide basis Swedish assistance made it possible to double the construction rate from 1965 to 1967, and further to quadruple this over the next two years (Warner, 1970). The Swedish aid was entirely based on planning and implementation through the Tanzanian bureaucracy. It was strongly focused on the technical aspects of rural water supplies for which Sweden provided substantial engineering manpower assistance. This consisted mainly of experts at central level and volunteers in the regions and districts (SIDA/PMO, 1983, Fig. 3.2).

During the 1970s several new donors started to operate in the sector and from the mid-70s to the early 1980s the Swedish share of total capital investments gradually fell to 30 per cent (SIDA, 1983b). At the end of this period Swedish support came under increasing attack – also from within SIDA itself. In 1979 the critics openly talked about the scandal of Swedish assistance to Tanzania's water sector: The large number of diesel schemes financed through Swedish aid were expensive and very difficult to operate and maintain; there was no participa-

---

[3] Only water activities are dealt with here (see Ch. 2).

Table 8. *Swedish Funding of the Lake Regions projects, 1983—1986*

Swedish kronor (×1000)	83/84	84/85	85/86
Expatriate Personnel[a]	7370	9930	12100
Promotion[b]	135	530	2100
Training	135	830	1500
Logistics & Infrastructure	2720	3600	4200
Water Supplies	6320	9300	9200
Planning Reserves	500	460	500
Total	17180	24650	29600

a Figures refer to budgets. Actual spending on expatriates has been reduced.
b Promotion of participation; training of village health workers, etc.
Sources: SIDA/PMO (1983, 1984a, 1984b)

tion of beneficiaries in the projects; the Tanzanian authorities did not have the capacity to plan and implement the Swedish aid to sector development efficiently; and numerous schemes did not function (*SIDA-Rapport*, 1979).

In the meantime Swedish aid to the rural water sector was slowly undergoing two major changes on the drawing boards. A gradual switch from country-wide assistance to the sector to support for the three Lake Regions only was being prepared; and this regional concentration of aid was to be based on careful planning by Swedish consultants. The plans were prepared by various consultants from 1975 to 1984. No implementation in the field based on those plans began until 1984. By that time the previous bias of Swedish assistance and plans towards the technical aspects of rural water supplies had been replaced by an approach with strong emphasis on the software side: participation, health, sanitation, management, and improvement of traditional water sources. The new approach has been baptized HESAWA (*he*alth, *sa*nitation, *wa*ter). It is to a large extent similar to the approach outlined in the new Swedish policy on aid to the water sector (SIDA, 1984). The funds allocated to the HESAWA project during the first three years of implementation are shown in Table 8. The Tanzanian authorities have approved the HESAWA approach in principle. To what extent the Tanzanian bureaucracy and the Swedish expatriates in the field, most of whom are used to a more technical approach, can and will adopt this new strategy is still an open question. Planning drastic changes in approach is far easier than implementing them since many of the ideas proposed have not been tried out in practice.

6.2. Planning

The RWMPs for the Lake Regions were prepared by Brokonsult - a Swedish branch of an American-owned company. The work was carried out from 1975 to 1977. Sixteen expatriates from nine different professions participated during various periods of the preparation. Five counterpart engineers from the Ministry were assigned – but for varying periods. The team operated from Mwanza, independently of the authorities in the three regions with whom no formal links were established. The consultants basically wrote up the plans themselves. There was no coordination with the concurrent rural integrated development planning team which worked in Mwanza from 1974 to 1976. This team was also funded by Sweden (see Ch. 6.10). It was the water sector proposal in the integrated plan that the World Bank decided to fund (see Ch. 7) – not the proposals in the RWMPs.

Although Brokonsult had wide experiences of water supplies from Africa, Asia and Latin America, the ambitious tasks specified in the TOR (written by SIDA and Brokonsult and approved by the Ministry) soon proved hard to fulfil. Large amounts of data, collected by air-borne geophysical surveys (carried out on the insistence of the Ministry) and by questionnaire surveys in 1500 villages, were never fully analysed. Neither was the write-up of the RWMPs ever completed. The consultant demanded additional fees to finalize the job, but SIDA refused to pay.[4]

According to the RWMPs, it would cost Tsh 349 mill. to implement the rural water supply part of the programme for the three regions to reach the 1991 goal (the plan is functional with respect to objectives). However, this did not include the cost of distribution systems from the source of water (normally a borehole) to the village. These costs were not estimated (Brokonsult, 1978a, Table 2). There were other peculiarities in the Brokonsult RWMP. They are discussed below.

No implementation was ever made on the basis of the RWMPs. Instead, a new Swedish consultant (VIAK) was called in to prepare a new implementation plan for the three regions. That work was carried out in 1980 and 1981. The VIAK team also worked independently of the recipient organizations. Their plans were discussed and approved in March 1982 and were to provide the framework for the objectives and direction of the Lake Region programme (SIDA/PMO, 1983, Appendix 3.2.1., p. 1). Shallow wells, participation, and management

---

[4] SIDA was also dissatisfied with the implementation approach proposed by Brokonsult. Interview with former SIDA official in May, 1984.

were the key words in this plan in which many ideas were based on the World Bank experiences in the Mwanza Project (Ch. 7).

However, the VIAK plans were to some extent overtaken by events. For in August 1982 IRA and WMCPU were asked to work out a more detailed plan of action for community participation, health education, and sanitation than was presented in the implementation plans. After extensive discussions from November 1983 to February 1984 the IRA/WMPCU report was approved in April 1984 (IRA/WMPCU, 1984, 3). It proposed an integrated approach, later called HESAWA (see Ch. 6.1) with strong emphasis on participation and partly inspired by the DANIDA project (see Ch. 8). However, before the HESAWA principles had been approved by the Swedish and Tanzanian authorities, SIDA had engaged a new consultant to coordinate the Swedish aid to the Lake Regions. HIFAB staff began to arrive in November 1983 and *their* terms of reference and professional background were based on the old VIAK plans of 1981. As discussed below, this did and does create problems in the implementation.

Whereas the RWMPs prepared by Brokonsult and the implementation plan prepared by VIAK were fairly control-oriented (especially the former), elements of a more adaptive planning and implementation approach have emerged since 1984. Planning and implementation (based on three-year rolling plans) are guided by a set of long-term goals on the improvement of water supplies, health and sanitation conditions, and the reduction of Tanzania's reliance on foreign aid. The short-term objectives are to build new low-cost schemes and to rehabilitate existing ones. But SIDA and the Tanzanian authorities have agreed to pursue a number of additional short-term objectives which are new in the rural water sector context. One is to improve traditional sources of water and not just concentrate on pumps and taps. Another set of objectives concerns the improvement of the capacity of recipient institutions and the beneficiaries to plan, implement and maintain the schemes built. Support for the introduction of new accounting procedures will also be given (SIDA/PMO, 1984b). Thus there is apparently not the same bias towards construction as in the other donor cases analysed here.

In this institutional development effort considerable use of short-term consultants has been made to analyse, to plan, to conduct seminars and to arrange for training in strategic planning, etc.[5] It is still too

---

[5] Three consultancies requested and financed by SIDA and three by HIFAB in 1984; three consultancies in 1985 (by SIDA).

early to judge how this influx of advisors will influence the work of the HIFAB expatriate staff and the Tanzanian staff in the recipient organizations involved.

6.3. Institutional Framework

The RWMPs of 1978 proposed that construction of rural water supplies in the three regions should be concentrated in an implementation unit under the Ministry, but taking its guidance on priorities from the RDDs and technical guidance from the RWEs. Technical assistance was to be given to this unit (Brokonsult, 1978a, 37). With respect to technical responsibilities this was a complete *reversal* of roles between RWE and Ministry: Tanzanian practice is that the former implement under the technical guidance of the latter. Brokonsult claimed (without evidence) that the Ministry would be more efficient in running the large-scale implementation called for by the 1991 goal. Furthermore, Brokonsult assigned the RWEs primarily to be in charge of operation and maintenance – a task which has always been shared with the DWEs.

With the arrival of the VIAK implementation plans such major institutional changes were quietly forgotten. VIAK basically recommended that existing organizational responsibilities be maintained and strengthened. Technical assistance should be fully integrated into the RWE offices. It was envisaged that expatriates could be withdrawn over a five-year period (VIAK, 1981, 65–67). However, it was proposed to establish zonal stores and procurement and a zonal training centre to cater for all three regions. This was in addition to existing Tanzanian structures.

VIAK (1981, Fig. 1.7 and 49) also proposed a new planning procedure. Based on a perspective plan geared to the 1991 goal, five-year action plans were to be worked out. In turn these were to be detailed in rolling three-year action programmes, linked to a new monitoring system.

Many of the VIAK proposals – as amended by IRA/WMPCU (1984) – now form the basis for the HESAWA programme (see above). A five-year technical assistance period is still regarded as sufficient, despite the ambitious institutional development aims that are pursued. A three-year planning cycle has also been maintained, and this is supplemented by one-year action programmes[6] (SIDA, 1983a). A joint Swedish/Tanzanian review mission visits the project each year and analyses past performance and approves future plans and budgets. Most major decisions appear to be made during this review.

The zonal coordination structure envisaged by VIAK and partly established in 1983–84 is now being abolished. The zonal coordinating role of the consultant, HIFAB, and the zonal Steering Committee with representatives from the regional authorities and central level are no longer significant. Instead, interregional coordination will be handled by the PMO with HIFAB in an advisory role only. Most coordination activities are now envisaged at the regional level and will be carried out by Regional Action Teams.[7] In the HESAWA review in 1985 it was further decided to channel HESAWA funds, services and activities through the districts rather than the regions as of April 1986. Thus the Swedish-funded project is the first one which has begun to take the consequences of the decentralization reform in 1984 (see Ch. 2.3).

Finally, it should be added that SIDA has much less day-to-day influence on the running of HESAWA than for example DANIDA has on the project it funds (Ch. 8.3). SIDA follows the project through a programme officer in the SIDA mission while DANIDA has established a donor-staffed steering unit. In the Lake Region project the consultancy firm, HIFAB, therefore plays a significant decision-making role.

6.4 Scheme Technology

According to the RWMP by Brokonsult (1978a, 13), main reliance should be placed on deep boreholes from which water should be extracted by diesel driven pumps. The consultant argued strongly against hand-pumped wells. They are more costly to *construct* on a per capita basis than diesel-pumped supplies, and "shallow wells will probably not be acceptable to villagers" (Brokonsult, 1978a, 13,35,37). However, in 1979 a World Bank hand-pump shallow well project for Mwanza was already under way (Ch. 7). The VIAK planners (1981, 25) also reached the opposite conclusion: wherever possible shallow wells or medium-deep wells with hand-pumps should be used. Deep boreholes should only be considered where hydrogeological conditions do not allow wells.

The HESAWA programme takes the concept of water supply improvement one step further. Apart from the construction of shallow

---

[6] Local funds from the government cannot be programmed on a three-year basis. Only one-year budgets are institutionalized.

[7] Members from RDD's office, water, health, and community development offices; CCM, and HIFAB.

and medium-deep wells, existing traditional sources must also be improved by capping springs, fencing water holes, etc. (IRA/WMPCU, 1984). This holistic approach to rural water supply is based on analyses of women's water use patterns at village level (see Ch. 2.1). How it will be carried out in practice has not yet been established. Pilot experiments began in 1986.

6.5. Service Level

The RWMPs were prepared from 1975 to 1977 during a period when the Party had declared that by 1981 all villages should be provided with one good source of water. It was a response to the massive movement into villages that took place around that time (Ch. 2.4). Brokonsult unquestionably based the RWMPs on the goal that one safe water source should be established within 5 km of every rural household by 1980, and a distribution system to bring clean, piped water within 400 metres of every household should be built by 1991.

To reach this target Brokonsult (1978a, 2–3) proposed a three-phased implementation schedule. In phase I (6 years) deep boreholes should be drilled in suitable locations for each village in each region. In phase II (4 years) diesel pumps should be fitted and water distributed to within 5 km of each village. In phase III the distribution system should be expanded to bring water points within easy reach of village households. While phases I and II should be financed by the government, according to Brokonsult, phase III should be financed by the villagers themselves. In this way their willingness to pay should determine the duration of phase III. It should be noted that the investments made during phases I and II would be unlikely to change women's water use patterns. But this was totally ignored in the plans.[8]

When VIAK took over from Brokonsult a two-phased approach was proposed. During phase I the government would construct hand-pumped wells within easy reach of all villages and with a capacity of 20 litres per person per day. Further increases in service level (including a change to piped supply) would have to be paid by the villagers themselves (VIAK, 1981, 7–12).

In the current HESAWA programme the number of wells constructed will be determined on the basis of the population of each village. However, the fitting of hand-pumps on to the open wells will depend on the willingness of villages to pay Tsh 3000 in cash for each

---

[8] Women's water use patterns are discussed in Chapter 2.1.

pump. A similar condition was introduced by the World Bank Project (Ch. 7). Furthermore, the project will assist villagers to improve traditional sources, since it is known that these "compete" with wells in women's choice of water source.

6.6 Priorities

Brokonsult (1978b, 344) gave the first priority to rehabilitation, operation and maintenance. If additional funds became available then new construction should start in "crisis" villages (dry season supply less than 10 litres per person per day). In Mwanza 230 such villages were identified (Brokonsult, 1978a, Table 3).

First priority to new schemes for "crisis villages" is the *only* major Brokonsult recommendation which can be found in the VIAK plan (1981, 8). This plan, however, recommends that rehabilitation, operation and maintenance improvements *and* construction of new schemes are pursued concurrently. In fact, a major activity in the VIAK proposal is the construction of new schemes (See Ch. 6.7).

The dual emphasis on rehabilitation and construction is maintained in the HESAWA programme. Mostly rehabilitation activities have been funded until now. The concept of "crisis villages" appears almost to have disappeared (SIDA/PMO, 1984b). New villages are selected by the regions based on unspecified criteria. The need to concentrate programme activities in specific areas (to reduce cost of transport, supervision, etc.) also makes adherence to implementation in crisis villages difficult.

6.7 Construction

Brokonsult (1978b, 458) envisaged that if sufficient resources were made available, new schemes should be constructed by an implementation unit under the Ministry – strengthened by expatriates (Ch. 6.3). This unit should implement phases I and II sequentially in each region (Ch. 6.5). By 1992 all three regions would have reached phase III. The 1991 goal *can* be reached. But, as Brokonsult (1978a, 12) stringently noted:

> Although it is most efficient to carry out the programme on the large scale . . ., if resources are not available then we recommend using what is available.

Also the VIAK plan (1981, 64) is based on the assumption that the 1991 goal can be reached "provided the main concepts on which the Implementation Plan are based are consistently and energetically pursued". Yearly production targets for wells in Mwanza reflected this optimism: 100, 250, and 340 respectively starting from 1981 (VIAK, 1981, Ch. 12.5). For comparison the World Bank Wells Project was struggling to reach *its* production target of 192 wells in 1981. It reached 89 after having been at it for two years (Ch. 7).

HESAWA production targets have been more subdued. Thus for 1985/86 a total of 156 sites are to be developed in Mwanza (SIDA/PMO, 1984b, Agreed Minutes, pg. 13). Nevertheless, in the preceding year construction progress had only been 50–60 per cent of the target (SIDA/PMO, 1984b, 6), amounting to 30 wells for the first nine months of 1984 (HESAWA, 1984, 2). In 1983/84 only 25 per cent of the target was reached with respect to shallow wells (SIDA/PMO, 1984a, Ch. 4.8). One of the problems is that village cash contributions for wells have not been forthcoming.[9]

6.8 Operation and Maintenance

Brokonsult (1978c, 163–180) envisaged a three-tier O & M set-up in the RWMPs. Main repairs were to be done by the RWE; corrective and preventive maintenance by DWEs; and day-to-day O & M by a villager employed and paid by the RWE. Villages themselves should have no active role in O & M. Financing of this system should in the long run come from the villages. In the meantime the government should continue to pay the bill. The change from one type of financing to the other is not described at all (Brokonsult, 1978a).

The VIAK plan (1981, Appendix 7.3) proposed a four-tier O & M set-up (a ward maintenance officer is included). The system is more decentralized than that proposed by Brokonsult. The district is the most important tier both for preventive and corrective maintenance. Villagers should select two pump attendants for training. Initially they should be involved in operation. Later they should also be in charge of some maintenance activities. The proposal assumes a 40–50 per cent increase in recurrent expenditure in 1981 compared to the regional budgets of the previous years (VIAK, 1981, 6 and 47). Villagers

---

[9] In 1983/84 no wells were constructed in Mara and Kagera. By the end of 1985 village cash contributions in Mwanza began to grow. The production rate was then increased to 12–13 wells per month.

should gradually pay the full cost of O & M but not for the first 5 years (VIAK, 1981, 30).

Some of these ideas also form the basis for the HESAWA programme, but they have not yet been worked out in detail. A short-term consultant has studied the existing O & M system (Book, 1984). In Mwanza it only functioned on an ad-hoc basis and was mainly run by the RWE. Based on this consultancy report, spare parts for one year's consumption were ordered. They have put most diesel-pumped schemes back in operational order. There is unfortunately no fuel to run them on a regular basis (SIDA/PMO, 1984b, 8). Despite this the regions – and some of the HIFAB staff advising them – have also recently requested funds for construction of new diesel schemes. However, all requests for funds for construction of new schemes are now rejected by SIDA.

Although the HESAWA planning procedures call for estimates of O & M costs for all new schemes requested, this was still not consistently done by all regions in 1984. And nobody yet has a clear picture of the total O & M cost of schemes already constructed. On this score the prime importance of O & M in the HESAWA short-term objectives has had limited impact so far among the Swedish and Tanzanian decision-makers in the regions. No serious work on transferring some of the financial and technical responsibility for O & M to the villages has been carried out yet. The short-term consultant did not elaborate on it. He stuck mostly to the technical issues (Book, 1984).

6.9 Participation

Participation was only accorded a limited role in the RWMPs. Brokonsult (1978a) recommended that (i) "village labour contributions may be more trouble than they are worth" (p. 38); (ii) "specifically we recommend against the village employing the pump attendants" (p.36); but (iii) "operations and maintenance costs should be shifted to the village by a systematic formula" (p. 38). No formula was provided.

The implementation plans by VIAK envisaged that the villagers actively participate in all aspects of water supply development – including design, construction, operation, and maintenance. The full cost of O & M and part of the cost of construction should gradually be covered by the villages. Likewise, health education and sanitation activities should be initiated in cooperation with the villages. Promotion units at regional and district levels should be established within the RWEs and DWEs offices. A Primary Health Committee at village level should also be established. Fairly detailed task descriptions for these units were given (VIAK, 1981, App. 7.4).

Participation is, in the HESAWA programme, *the* important activity. It is advocated in all activities (water, sanitation, health; IRA/WMPCU, 1984). The main features are to some extent based on proposals from the DANIDA project (Ch. 8) and from proposals by PMO/IRC (Ch. 5). They will not be further discussed here. The strong emphasis on participation, water improvements, health education, and sanitation activities is only slowly being reflected in the composition of the donor-funded HIFAB staff. Of the 16 expatriates recruited (or under recruitment) by the end of 1984, only one was hired to work on participation activities (SIDA/PMO, 1984a, 8). The rest were engineers (5), mechanics (3), accountants (3,) training specialists (1), and procurement officers (1). By 1986 a health advisor and two Tanzanian promotion officers had been added to the SIDA hired staff.

6.10 Coordination

It is interesting to note that the RWMP exercise carried out in Mwanza from 1975 to 1977 was not coordinated in any way with the rural integrated development planning exercise (Ridep) carried out from 1974 to 1976. Both exercises were funded by SIDA. The two plans differ significantly, for instance with respect to technology choice and participation in rural water supply activities. Yet both plans were "approved" by the Tanzanian authorities and this raises interesting questions about the status of donor-funded planning (see Ch. 2.2). But since the World Bank was prepared to fund a water component in the Ridep proposal, the RWMPs were in effect disregarded (Ch. 7).

Turning to the content of the RWMP, Brokonsult (1978a, 12) called for "prompt" decisions on the "method of establishment of the administrative apparatus necessary to support the execution of a large, technically complicated project". But no proposals or alternatives were put forward by the consultant.

VIAK (1981, 2) argued that the planning process proposed is the main tool for coordination (5-year perspective plan and 3-year rolling action plan combined with annual reviews and monitoring and evaluation; see Ch. 6.11). Basically, the proposed programme is seen as a one-sector activity, planned and implemented by the water authorities. Therefore coordination with other sector activities was not considered in much detail – although some overlap with health and community agencies was foreseen (VIAK, App. 7.4, p. 1–2).

In the HESAWA programme the coordination problems are very real, because the aim is to coordinate the activities of three Ministries (and three regional functional offices) at the village level: water (MAJI); health/sanitation (AFYA); and community development (MAENDELEO). The answer so far has been to establish a number of committees at zonal and regional level – and eventually at district and village level.[10] The regional committees have tended to grow in membership and have been very cumbersome to work with and within.

An equally serious problem is who should coordinate who. MAJI plays a prominent role in the HESAWA programme as reflected in the resource allocations (Table 8). Yet AFYA and PMO activities are equally important in the HESAWA concept. "Crucial inputs from MAENDELEO and AFYA may not be forthcoming if they feel unduly dominated by MAJI and deprived of their fair share of the overall project resources" (Nordberg, 1985, 18). In the last programme review it was recommended that the three agencies plan their activities together (SIDA/PMO, 1984b, Ch. 2). Hitherto this has not happened and there is no tradition of such joint planning in Tanzania. Also in this respect the HESAWA concept is challenging the working relationships between recipient institutions. The attempt by SIDA and the PMO to place the latter as the main coordinating agency has, furthermore, challenged the de facto control by the water authorities over rural water sector activities.

6.11. Monitoring and Evaluation

There are no proposals on this issue in the RWMPs. In the VIAK plan (1981, App. 12.1) yearly reviews based on a set of simple monitoring forms were proposed. They allow a comparison of planned versus actual performance to be made.

This idea has been much elaborated in the HESAWA programme by a short-term consultant (Samset and Stokkeland Consulting, 1984). However, their detailed proposals are not designed for a multi-agency, participatory grassroot programme as HESAWA is supposed to be. Thus the proposed traditional project management monitoring and evaluation system (i) envisages an upward flow of information to

[10] The committee at zonal level has now been abolished, see also Chapter 6.3.

the assumed decision-makers at the regional level; (ii) proposes to quantify inputs and output of all activities (including participation and health education); and (iii) assumes that only one (MAJI) of the three agencies (AFYA and MAENDELEO) involved, collects information. However, these proposals are not being implemented with respect to the non-technical field activities (participation, health, sanitation). No doubt programmes of the HESAWA type stretch conventional monitoring and evaluation approaches to the limit!

6.12. Key Observations

The RWMPs prepared in the late 1970s are a good example of a control-oriented long-term plan based on intensive data collection and analyses and prepared outside the recipient institutional framework for these activities. It had very little impact on subsequent implementation.

The VIAK plan which followed was also a long-term control oriented plan. Much fewer resources were spent preparing it and the planners did benefit from the implementation experiences of the World Bank funded project in Mwanza (Ch. 7). Both the proposals on the shallow wells technology itself and on village cash contribution are taken from this World Bank project. The VIAK plan provided the basis for the start of the implementation at the end of 1983. But the framework for implementation since then has been strongly influenced by the proposals by two social scientists from IRA and WMPCU and the new SIDA water sector policies issued in 1984.

The intentions in the HESAWA project present a clear break with previous Swedish efforts and some of them are new in Tanzania: (i) the emphasis on improvement of traditional sources with simple means requiring use of very low cost technology; (ii) the integration of health, sanitation and water activities requiring interagency cooperation and coordination; (iii) the attempts to place the PMO – not the water authorities – as the lead agency in these efforts; (iv) attempts to decentralize activities to the district level, requiring institutional realignments; and (v) the emphasis on strengthening the capacity of local institutions to carry out the various activities requiring a strong emphasis on training.

Practices at the end of 1985 tended to differ – sometimes significantly – from these intentions. Given the age of the project at that time, this is not surprising. The project seeks to break new ground. This creates many practical problems which can only be solved by trial-and-error and which take time.

In the end the outcome of the project will depend on the capacity and willingness of local institutions from village to national level to participate in this process, and on the ability of Swedish staff to assist. The project faces major problems here as noted by one consultant (Nordberg, 1985, 15) in early 1985 and confirmed by another (Arnzen, 1984):

> There is a superficial knowledge and limited understanding of the crucial elements of the HESAWA programme, even among people involved in implementing the programme. On some parts there may be a certain reluctance to recognize the proposed changes in emphasis and approach.

The question of bureaucratic incentives appear to be important here. What means will, or can, the HESAWA project use to motivate the various agencies to implement activities that differ significantly from past approaches? It is clear that the five-year time horizon within which the project currently operates is far too short. A long-term strategy is needed to provide the guidance for the learning process that the project has just embarked on. Suggestions for such a framework are discussed in Chapter 11.

7. The Project Cycle "Approach": The World Bank in Mwanza Region

World Bank assistance to the rural water sector in Mwanza region was initiated in 1977 and ended in 1984. This assistance was part of an integrated project titled the Mwanza Rural Development Project (MRDP). The planning and implementation of the water component of this project is interesting because:

– only project planning – not sector planning – was carried out;

– the RWMP, completed by the Swedes in 1977, had almost no impact on World Bank proposals;

– throughout the project cycle, standard World Bank approaches to appraisal, planning and evaluation have been used (described, for instance, in Baum (1982));

– a participatory approach to water supply development was planned, based on socio-economic studies;

– an integration of planning and implementation into existing local instructions was envisaged and planned;

The World Bank thus deliberately attempted to avoid some of the major problems of the turn-key approaches described in Chapters 4 and 5.

7.1. Description of the Project, 1977–1984

The World Bank planners were inspired by the Dutch shallow wells project in Shinyanga (see Ch. 5). They saw hand-pumped wells as a least-cost solution to the rural water supplies problem and also as a suitable technology for village self-help projects.

Table 9. *World Bank funding and output of the Mwanza project, 1979–1983*

	Mill.Tsh	People 'served'[a]
Construction of wells	14.6[b]	68,000
Rehabilitations of piped supplies	0	–
Operation and maintenance	0.2	–

a Assumes that all 228 constructed wells are functioning
b Excludes RWE and village contribution
Sources: PEU (1984, Tables 1, 2 and 9)[1]

In 1977, when the World Bank presented its project proposal to the regional authorities, SIDA also presented its findings from the Water Master Plan Study. The Swedes concluded that hand-pumped wells were inappropriate and too expensive for Mwanza region (see Ch. 6). The Regional Water Engineer was – in his own words – "a bit surprised" by the two contradicting opinions. But the World Bank was willing to put up approximately Tsh 24 mill. for wells over a five-year period and that settled the matter.

The objectives of the water component of the World Bank financed MRDP were (i) to use participation as a "mechanism" through which to identify, prepare, construct and finance all water supply systems installed by the project; (ii) to create a shallow well construction and maintenance division within the RWE office; and (iii) to rehabilitate some piped schemes (World Bank 1977b, 4).

Table 9 shows the resources spent by the water component and the additional number of people served during the period 1979–83. For most of this time three to four expatriates were attached to the shallow well unit. Both the project manager and the geologist had previous experience from Shinyanga and both had Tanzanian counterparts through most of the period.

World Bank financing of the integrated rural development project stopped in 1983 and subsequently resources for water supply development dried up. The RWE continued to construct some wells but at a reduced level. By 1984 Sweden was back in the region again and prepared to continue shallow wells construction activities (see Ch. 6).

[1] Project Evaluation Unit established and staffed by the World Bank to evaluate all components in the MRDP.

7.2. Planning

Two documents provided the guidelines for the water component of the MRDP. One described the background, concepts and procedures for a village self-help mechanism through which villages and district/regional authorities could cooperate to establish non-self-financing and self-financing projects in the villages (World Bank, 1977c). It was, among other things, based on a fairly detailed "socio-cultural profile" in which the historical roots of self-help activities among the Wasukuma were traced (World Bank, 1977a)[2]. No other donor in the water sector has undertaken a similar sociological study. The other planning document set out in some detail the technical aspects of the water component (World Bank, 1977b). The plan and project period was five years, during which a gradual growth in activities was envisaged. What should happen after the first five years was not described in the World Bank plans. The 1991 goal was not incorporated in the plan at all. Neither was there much reference to the RWMPs prepared by Brokonsult (Ch. 6).

The two documents together provide a concise, mercifully short and yet fairly detailed description of the background, objectives, scope, procedures and actions needed to set the water component in motion. They also emphasize the non-technical aspects of rural water supplies such as management, organizational issues, socio-cultural conditions and participation. From a technical point of view the project documents are better and more comprehensive (despite their brevity) than similar documents prepared by other donors for other regions. The question is: does better comprehensive pre-planning of projects lead to better plan compliance during implementation? The answer to that question is not affirmative, as shown below.

7.3. Institutional Framework

Four organizational changes of relevance to the water component were envisaged in the project documents. First, a wells construction division was to be established and integrated into the RWE's office (World Bank, 1977b, 5). The unit was actually established, but it "had considerable autonomy with its own budget and was never fully incorporated" within the organization of the RWE (PEU, 1984a, 21). Second, a well maintenance division was envisaged in the project documents, but it was never established (PEU, 1984a, 12 and below).

---

[2] The Wasukuma is the largest tribe in Mwanza region.

The third organizational change was the establishment of a village self-help project unit (VSHP) within the regional planning office.[3] It was to create procedures and structures for "substantial village participation in the determination of well priorities, in project identification, formulation, execution . . . and financing" (World Bank, 1977c, 4). Finally, a village self-help programme board was to be established to formulate policy, provide overall guidance and grant final approval to, for instance, water projects. This board was to consist of civil servants, politicians and World Bank staff. It was to operate outside existing local coordinating bodies such as the Regional Management Team and the Regional Development Committee (World Bank, 1977c, 12–13). Thus the MRDP had its own parallel decision-making structure[4].

7.4. Scheme Technology
The well techniques used in the Mwanza project were those developed in Shinyanga and later in Morogoro (see Ch. 5), as envisaged in the project plan.

7.5. Service Level
According to the project plan each well was to serve a maximum of 300 people, and each village was to be provided with 7 wells on average (World Bank, 1977b, 4). However, political pressure to provide wells to as many villages as possible and the demand that villages contribute Tsh 6000 in cash for each well (see below) caused the 228 wells constructed by the project to be spread between 200 different villages (PEU, 1984a, 3). This means that on average the number of people obtaining water from one well is in theory between 360 and 2400 (PEU, 1984b, 3). In practice, many households in the villages provided with a hand-pumped well will continue to use traditional sources rather than queue up at a well.

7.6. Priorities
In the Water Master Plan prepared through Swedish aid and completed in 1977, a number of high-need "crisis" villages had been

---
[3] A staff function under the RDD (see Ch. 2.3).
[4] Both the self-help unit and the board *were* set up. Their activities are described below.

identified (see Ch. 6). These crisis villages were to get the highest priority in the World Bank funded project as well (World Bank, 1977b, 7). However, the village cash-payment condition conflicted in part with the need criteria and the local politics of the selected villages. As a result, the RWMP priority criteria had little impact on actual implementation.

7.7. Construction

World Bank planners, concerned with cost-recovery and with ways of committing villages to water projects, proposed that villages should contribute 25 per cent (Tsh 6000) of the construction cost of each hand-pumped well established. This directly contradicted Tanzanian policies at the time, but was nevertheless approved by the authorities. Obviously, it was unknown to the planners *how* the users would react to this new approach. Despite this major uncertainty the World Bank planners wrote exact yearly production targets into their plans (0, 24, 108, 192, 216 for the 5-year period). The whole programme and cost calculations are based on these figures. Furthermore, to introduce shallow wells in the region, the project design envisaged that a total of 20 demonstration wells be constructed free of charge for the villages. These wells, together with a coordinated publicity campaign organized by the VSHP, were expected to generate requests from villages for additional wells for which they were to contribute in cash (World Bank, 1977b).

What happened? By July 1980 (12 months into the project period) no requests from villages had been received and it was agreed to increase to 50 the number of free demonstration wells. By December 1980 there were still very few requests. At the end of the project period only 228 out of 540 wells were completed, but just 22 per cent of these were actually paid for by the villages according to the initial conditions. Instead almost three out of four wells were funded by the World Bank, the districts, the region, the missions or by charitable organizations (PEU, 1984a). The project did not have the patience to wait for the village demand to build up to match the planners' assumptions. To this should be added the point that the plans for production of new wells were "over-optimistic and did not take into account the logistics and administration that would be required" (PEU, 1984a, 6).

As a result of all this, less than 50 per cent of the planned physical

---

[5] Villagers were, however, working with the construction crews, but as paid labourers.

target was reached and this obviously also spoiled the estimated costs. Wells became much more expensive to construct than anticipated. Instead of the Tsh 24,000 originally estimated the actual cost turned out to be around Tsh 70,000 (PEU, 1984a, 15).

An ambitious programme of rehabilitation of 17 piped supply schemes was also envisaged. For the first single-village scheme rehabilitation, Tsh 160,000 was provided by the project, with Tsh 40,000 (25 per cent) expected as village contributions. The selected village was unwilling or unable to produce the cash. In the end the RWE rehabilitated the scheme without waiting for village funds. With this precedent no further attempts were made to rehabilitate piped schemes. A whole project component was given up.

7.8. Operation and Maintenance

It was proposed in the project design that assistance would be provided to the RWE to establish a maintenance capability within the existing organization. This was to consist of two mobile teams, one for general maintenance and one for well repairs. The villages were supposed to bear a significant part of maintenance costs (Tsh 600 per well per year) and to sign a contract with the VSHP to that effect. In addition, villages were supposed to nominate two people per well to receive training as hand-pump attendants:

> The two maintenance teams were never established. During the lifetime of the project, efforts were concentrated on . . . construction. No contractual agreements were made with villages . . . and there have been no village contributions towards maintenance . . . . The training programmes were never organized and no pump attendants were appointed . . . . The [maintenance] aspect of the programme appears to have been totally forgotten about (PEU, 1984a, 12–13).

In the absence of any organized maintenance the RWE has, since the start of the World Bank project, done repairs on an ad hoc basis under severe constraints of fuel, vehicles and funds, made worse by the fact that single wells are scattered in some 200 villages.

7.9. Participation

In the project write-up the full participation of villages in all project activities was listed as the prime objective (World Bank, 1977b). The feasibility of this approach was based on sociological studies in which it was concluded that:

> the systematic promotion of self-help developmental activities fits well into Sukuma structure and behaviour. Traditional customs can be relied upon to promote the new approach, since the compatibility between VSHP and the traditional Sukuma culture offers a propitious context (World Bank, 1977c, App. 1, 8).

However, the planners warned that the Government of Tanzania and the World Bank

> should refrain from equating success with quantitative criteria.... Genuine achievement of [the participatory] objectives should be the principal criterion of success, even if in practice this required a substantial reduction in the size of the programme (World Bank, 1977c, 9).

There were three key elements in the proposals aimed at participation. One was the VHSP unit in the RDD's office already discussed in Chapter 7.3. It was to be the coordinating and planning body for the exercise. Another element was the ward-level community development officers, who were to act as extension agents between the authorities and the villages. The third element was the set of elaborate procedures laid down in the project design papers (World Bank, 1977c). All was apparently set.

Unfortunately, villagers refused to participate under the conditions stipulated and/or with the staff operating in the field. As already discussed, villages contributed cash only slowly, and the project staff could not or would not involve villagers in planning, construction, and maintenance. It did not help that the VSHP unit soon proved to be understaffed and that the work of the VSHP unit and the wells construction unit was not coordinated (PEU, 1984a, 6). Furthermore, the project manager for the wells construction unit was a professed non-believer in participation (Ringelberg, 1980). In the end the construction unit ended up doing everything itself without involving villagers so as not to "severely handicap the implementation of the project" (PEU, 1984a, 7). Apparently the financer and the recipient did not object to this abandonment of the prime objective of the project!

7.10. Coordination
The project design incorporated the project into existing local organizations and established a coordinating body for self-help activities. However, the wells construction unit had "no close liaison with the VHSP component, which did not have adequate staff" (PEU, 1984a, 23). Each component tended to operate on its own.

The RWMPs prepared by Swedish consultants from 1975 to 1977 and the regional implementation plans prepared by another Swedish consultant in 1980–81 had no discernible impact on World Bank activities. Each donor proceeded independently of the other and Tanzanian authorities approved what each donor did (see Ch. 6).

7.11. Monitoring and Evaluation
A planning and evaluation unit was set up in the Prime Minister's Office in Mwanza to monitor and evaluate the different components of the Mwanza Rural Development Project. The monitoring (about which no information is available) did not prevent the water component from being completely diverted from its main objective: to base rural water supplies on active participation of the users. The excellent ex-post evaluation produced by the unit has pointed out this and other divergencies between plan and implementation. The implications for planning and implementation of rural water projects are, however, not discussed in the evaluations, and the project was abandoned before it could benefit from the evaluation.

7.12. Key Observations
The five-year plan produced by the World Bank for the water component of the Mwanza Regional Development project differs in several ways from other donor-funded plans. Thus the five-year plan for the water component is normative in that it is based on a scrutiny of means and ends. One clear result of this was the abandoning in the project of water as a free public service. The five-year plan is clearly blueprint-oriented. It is prepared prior to implementation by planners working outside the recipient planning machinery. It provides comprehensive, detailed specifications of project inputs and expected outcomes for the entire five-year period. However, no alternatives were presented in the plans.

This case study illustrates that even if the head is clear the hands will not necessarily follow. The five-year plan was both clearly formulated and well balanced between technical and non-technical proposals. Yet, the project evaluations reveal major differences between what was planned and what was actually implemented.

It is clear that the planned yearly production target for wells has been the driving force in project development. It has displaced other major objectives written into the five-year plans (participation, training, maintenance activities, etc.) Even the well-known World Bank insistence on cost-recovery in basic services programmes succumbed to the production pressure. It is clear that this particular goal displacement did not met with much complaint from the Tanzanian authorities.

Lele (1975, 129), investigating World Bank activities in Africa during the early 1970s, has described the planning mechanisms that lead to such displacements very well. Her explanations bear extensive quotation:

> However, there is an additional characteristic of the integrated programs central to their design and performance. Because there are no easy objective criteria by which to judge such accomplishments as the training of manpower or the development of administrative abilities, investment in these integrated programs tends to be judged primarily by the criterion of an acceptable internal rate of return, calculated on the basis of quantifiable production targets. These targets ... do not explicitly take into account some of the crucial complementarities in realizing production objectives – namely, the supply of trained manpower ... the effectiveness of administrative procedures, or the existence of other physical infrastructure such as roads .... Thus, the larger the proportion of expenditure in a given project on these latter types of components, the greater appears to be the need for ambitious production targets to carry the burden of these indirectly productive activities, so that the project can be acceptable in terms of its internal rate of return. This may necessitate setting ambitious production targets in the short run, introducing a contradiction in the project design from the very outset, for such targets may distract the attention of the project authorities from acquiring and training competent indigenous staff, from evolving administrative procedures that will last long past the stage of donor financing, and from developing effective working relations with the normal administrative structure.

The very same mechanisms have been at work in the water component of the World Bank project in Mwanza.

8. The Village-by-Village "Approach": The Danes in Iringa, Mbeya and Ruvuma Regions

From 1970 to 1980 Danish aid to rural and urban water supplies accounted for 6 per cent of the total aid budget. Since the launching of the Water and Sanitation Decade this percentage has grown, and for 1984 and 1985 it will be 12–15 per cent of Danish bilateral assistance (WHO, 1985, 52). This aid has been given to numerous countries in Asia and Africa. Active and substantial Danish involvement in the Tanzanian rural water sector started in 1979 with support for the preparation of three RWMPs and the construction of a number of water schemes in Iringa, Mbeya and Ruvuma regions. This support has continued since then. At present, approximately 50 mill. D.kr. yearly are allocated to water projects in the three regions.

The significant features of the Danish involvement are that

- the RWMPs were prepared by a combined team of technical and socio-economic planners;

- the RWMPs contain specific procedures for village participation in water projects (similar procedures for the technical activities were not prepared);

- during the preparation of the RWMPs, the RWEs constructed new schemes with Danish funds and materials; the socio-economic – but not the technical planners – were directly involved in this implementation;

- the design and priority criteria, and the procedures for participation stipulated in the RWMPs are now being adhered to in the implementation;

- the overall management responsibilities on behalf of DANIDA for the RWMP preparation and the subsequent implementation in the three regions rests with a donor-staffed, Dar es Salaam based, Steering Unit (SU);

– the donor control over the implementation process has increased markedly since 1980; executive powers have gradually been shifted from the RWE organization to a technical assistance team operating within each RWE office.

The details of the Danish involvement follow below.

8.1. Description of the Project, 1979–1985

In April 1979 DANIDA's board agreed to finance the preparation of three RWMPs and the construction of some new water schemes in Iringa, Mbeya and Ruvuma regions. A total of 74 mill. D.kr. was allocated for this first phase (1979–1982). After an interim period from March 1982 to the end of 1983 a new government-to-government agreement for a second phase (1984–1988) was signed, according to which 249 mill. D.kr. was provided for the implementation of parts of the RWMP proposals – notably the construction of new supplies for some 300 high priority villages. Table 10 indicates the funding and outputs of the DANIDA assistance.

The initial negotiations in 1979 about the RWMP preparations illustrate very well the particular donor-recipient relationship which is typical of the Tanzanian water sector. In principle, Danish grant aid is untied. Based on a list prepared by the Ministry of Water, DANIDA therefore decided to invite five consultancy companies to submit tenders for the preparation of the RWMPs. Two of these were Swedish. The selection of a winner proved very difficult. An independent evaluation of all tenders by a Danish consultant came to the conclusion that one Danish proposal (preferred by DANIDA) and one Swedish proposal (preferred by MAJI) were equally good. In the end – after active Danish lobbying – the contract for the preparation of the RWMPs was given to the Danish consortium of consultancy companies, CCKK, despite strong Tanzanian objections. "They were pushed down our throats" as one high-ranking Tanzanian, who was involved in the selection, remarked.[1] There have not, however, been complaints about the actual general performance of the Danish consultants.

[1] Based on interviews with MAJI and DANIDA staff.

Table 10. *Danish Funding and Output of the Iringa, Mbeya and Ruvuma Projects, 1979—1988*

Activity	Period	Funds (mill. D.kr.)	People served by Sept. 30, 1985
Preparation of RWMPs	1980—83	37.0[a]	—
Implementation of new schemes			
Phase 1	1980—83	37.0[b]	69.000[b]
Phase 2	1984—85	88.5[c]	70.000[b]
Phase 2	1986—88	147.0[c]	

a) DANIDA (1984a)
b) DANIDA (1984b, annex 4) and DANIDA (1985b)
c) DANIDA (1986, 11)

However, it was more than a Danish preference for a Danish company, and a Tanzanian preference for a Swedish company which had already prepared four RWMPs in the country and had good working relations with local authorities, that separated the two parties. The Swedish consultant proposed to collect a substantial part of the information on water resources by remote sensing. CCKK proposed that such information should be collected mainly by visits to all 1509 villages in the three regions. The Tanzanians insisted that this simply could not be done within the allocated time and that there was no need to collect such disaggregated data for a RWMP anyway. DANIDA claimed that the village-by-village approach to information collection and plan preparation would make the RWMPs much more implementation- and village-oriented than those made previously. This claim is further discussed below.

It was also upon Danish insistence that a three-year socio-economic study (SEC) was conducted as part of the RWMP preparation. The study focused on socio-economic base-line studies and on participation and health issues. It was conducted by a joint team of five researchers from BRALUP (later IRA) and CDR — of which three were residents in each of the respective regions.

While the RWMPs were being prepared the RWEs constructed a number of new schemes with Danish assistance. Around 69,000 people were provided with access to improved water in this way from 1980 to 1983, as shown in Table 10. The main RWMP consultant, CCKK, was not involved in this activity. Danish assistance consisted of funding, importing of construction material and some limited support for transport. Design and supervision assistance was provided by one technical advisor in the DANIDA-staffed Steering Unit in Dar es Salaam. The SEC became increasingly involved in experiments with village participation during this period, especially during the last year of phase 1.

In September 1983 Denmark undertook to finance water supplies for some 300 high-priority villages over a five-year period and to support some pilot projects in health education and sanitation in the three regions. This second phase of the project was allocated 249 mill. D.kr. By the end of 1985 a total of 70,000 people had been supplied during phase 2, as shown in Table 10. Compared to phase 1 both the implementation rate and the number of donor-employed expatriates and Tanzanians have grown significantly, as will be discussed below.

8.2. Planning

Three types of plans are used in the Danish-funded project: RWMPs, three-year rolling implementation plans, and detailed one-year plans.

The RWMPs for all three regions are contained in 40 volumes. Two interregional volumes have been prepared by SEC, the rest by CCKK. More than 300 man-months of professional staff time was used by CCKK to prepare its part of the RWMP, supplemented with some 150 man-months of professional Tanzanian counterpart time. The SEC used some 120 man-months for the RWMP preparation. Thus around 50 man-years of professional work went into the RWMPs.

According to the RWMPs, a total of 1300 mill. D.kr. would be needed to supply all 1509 villages in the three regions with an appropriate water supply by 1991. Some 630 of these villages were identified as high-priority villages. They could be supplied at the cost of around 400 mill. D.kr. The yearly cost of running and maintaining the schemes of all villages in 1991 would be 92 mill. D.kr.[2]

The RWMPs are functional plans in that they take the 1991 goal as given (see Ch. 3.2). Although the RWMP planners soon realized that the goal was unrealistic, they felt that the consultancy contracts and the terms of reference neither allowed nor required that more realistic goals be established. Moreover, this top-down planning goal contradicted the other basic objective of the RWMPs: that villages should participate actively in the planning, construction and maintenance of

[2] 1981 price level.

schemes. No scheme should be built without full agreement by the village. The dichotomy between these two planning approaches was left unsolved in the RWMPs.

The terms of reference for CCKK stated that the objectives of the RWMPs were "to provide the government of Tanzania with firm recommendations for the development of water resources . . . over the period 1981–1991, and in brief outline for an additional 10 years" (CCKK, 1982a, p. 2.1). Clearly this requires the consultants to produce a blueprint plan of the type discussed in Chapter 3.4. CCKK's part of the RWMPs does indeed contain a large number of firm recommendations. It contains, for example, specific proposals on the water source, scheme technology and cost of water supplies for each and every one of the 1509 villages in the three regions. These proposals are the backbone of the CCKK-produced part of the RWMP. No other RWMPs present such detailed and specific proposals.

However, on a number of key issues the RWMP is much less firm or sometimes almost silent. Training, for example, is dealt with in only a couple of pages of general descriptions and proposals (CCKK, 1982d, Ch. 11.8), although a five-fold increase in skilled labour for construction over a two-year period is envisaged in the plan (CCKK, 1982d, Table 10.21). Proposals on organization and management are also made in very general terms such as:

> to apply strategies and guidelines which will result in a precise, quick and realistic development of those parts of the organizational system which are vital to fulfilment of the overall objective of total effort . . . (CCKK, 1982d, p.11.9)

It is not easy to operationalize such recommendations. The proposals made by the SEC in volume 12 of the RWMP suffer the same problem (BRALUP/CDR, 1982). They were mainly based on questionnaire analyses and information collected through village inventories (see below). Only during the third year of work, when the SEC got actively involved in actual implementation of water schemes at village level, it was able to write a set of operational procedures for village participation (IRA/CDR, 1983). Unfortunately, CCKK did not get the same chance to combine plan-preparation with direct implementation experience.

A final point on the RWMP exercise concerns the approaches to information gathering and analyses. Much effort was spent on village visits in which a wide range of information was collected by CCKK.

The questionnaire prepared by CCKK and the SEC group in cooperation contained some 250 items. The planners had the familiar problems distinguishing what is "nice" to know from what is "necessary" to know. But there are more serious problems with the approach.

One purpose of the village visits was to collect information to establish the best water source, the appropriate scheme technology, and the cost of a water supply for each village. Such work had already been done by MAJI in a limited number of villages. Comparisons between CCKK and MAJI proposals are therefore possible. They indicate that "in very few cases have CCKK proposals differed from MAJI schemes either existing, proposed or under construction" (CCKK, 1981). This illustrates that MAJI is quite good at selecting the appropriate water source and scheme technology for rural water schemes. The need for technical assistance for this task has therefore been considerably overestimated.

A second problem concerns the reliability of the information collected. On average, a CCKK team spent around two to three hours in each village during which time most village information was collected through observation, sampling and interviews. Despite the very impressive efforts put into the tasks, experiences from the subsequent implementation period show that (i) village maps prepared during the visits are inaccurate and only of limited value even for preliminary design; (ii) the yield of low-flow streams – which are important sources of water for village water supplies – cannot be established with sufficient accuracy by the few spot measurements which were possible during the RWMP period;[3] and (iii) the actual potential for hand-pumped wells established through trial-and-error drilling during implementation is significantly higher than indicated by the RWMP. The implications of these findings are now being taken into account. Information collection is geared to the implementation plans. When new villages are included in the three-year implementation plans, they are also included in an information updating system. In this way important information is collected more carefully and over a longer time than was possible during the RWMP. No general updating of the RWMP data is attempted.

The SEC also faced serious methodological problems in their own questionnaire surveys.[4] It was, for instance, of crucial importance to

---

[3] Low flow gauging in 1984 indicates that approx. 50 per cent of the measured high priority sources yielded too little water to supply the year 2006 population they were intended to serve (DANIDA, 1985b, 42).

analyse the ability and willingness of villagers to contribute in cash or kind to the operation and maintenance of water schemes. Consequently a number of hypothetical questions on those issues were asked. (How many shillings would you contribute if . . . ? etc.). Respondents were both earnest and helpful in their answers and these were dutifully analysed (BRALUP/CDR, 1982, Table 6.9). Unfortunately, the concept of payment for water is largely unknown in rural Tanzania (it was also against official policy at the time). And to many respondents even the idea of getting a water scheme must (rightly) have seemed an unreal possibility. The fact remains that no social science methodology exists that will provide reliable and quantifiable answers to such hypothetical questions (Saunders and Watford, 1976, Ch. 7). The answers can only be found by trial and error, as slowly realized by the SEC during the third year of their work (see below).

As scheduled, CCKK and SEC completed their parts of the RWMPs in March 1982 and May 1983 respectively. In mid-1983 the RWMPs were approved with minor adjustments and a new government-to-government agreement for a second phase was signed by the end of that year. In the agreement it was specified that Danish assistance should be based on three-year rolling implementation plans. The preparation of these plans is very interesting from a planning point of view.

First, it has proved difficult to transform the long-term RWMP into a three-year implementation plan. Taken at face value the government-to-government agreement is clear enough. It stipulates that Danish assistance "will be carried out on the basis of the proposals and recommendations contained" in the RWMPs, especially as regards priority and design criteria and village participation. But this statement hides substantial ambiguities. Although the RWMPs are approved by both governments this does not imply that they agree to the proposals in them (see Ch. 2.2). Many of the recommendations on participation, for example, concerned changes in operation and maintenance policies that were highly controversial and which would need Party and Parliament approval before they could be implemented. Such policy approvals have not yet been made (see Ch. 2.4). Furthermore, the Danish government was unlikely to finance

[4] This survey was limited to some 20 villages in 3 or 4 clusters in various agro-ecological zones in each of the three regions. Within each village some 30 households were randomly selected for interviews. Furthermore, in-depth surveys (open-ended questions with non-quantifiable answers) were conducted in one village in each cluster.

water for all by 1991, the policy on which the RWMP was based. It is therefore not surprising that the *least* controversial parts of the RWMP proposals have ended up in the first implementation plans. And there is little doubt that this implementation plan – written by the DANIDA Steering Unit – was as strongly influenced by SU's own experience from involvement in implementation during phase 1 as by the proposals in the RWMP.

Second, the objectives of the Danish assistance to implementation are not clearly stated in the implementation plans. Such statements are also lacking in the RWMPs[5] and in the government-to-government agreements. Apart from production-oriented objectives written into the one- and three-year plans the management of project activities proceeds without any explicit development objectives. For example, operational guidelines are lacking for such crucial issues as: the balance between support for new schemes, rehabilitation and operation and maintenance; the relative importance of health improvements in project activities; the role of institutional development of local organizations in project activities, etc. More about this below.

Third, the resource allocations on which the three-year plans are based, have proved very difficult to predict. On the one hand, the government-to-government agreement does not stipulate the size of the Tanzanian contribution to the activities.[6] The one-year Tanzanian budget cycle would, in any case, prevent three-year Tanzanian commitments (see Ch. 2.5). On the other hand, the Danish allocations have proved to be quite variable and unpredictable. According to the government-to-government agreement of December 1983, 249 mill. D.kr. was to be allocated to project activities over a five-year period. The first three-year implementation plan – also prepared in December 1983 – was therefore based on allocations of 42,53 and 71 mill. D.kr. respectively for the period 1984 to 1986. Yet only two months later the Danish delegation to the annual Tanzanian-Danish aid negotiations argued that the precarious economic situation in Tanzania did not justify this activity level. Only 30 mill. D.kr. was therefore approved for 1984. In 1985 and 1986 the allocations have again been increased to around 50 mill. D.kr. while economic conditions in Tanzania steadily

---

[5] However, a number of explicit objectives have been stated by IRA/CDR (1983), but only a few of these have actually been used in subsequent implementation.

[6] The government-to-government agreement stipulates that "villages are supposed in principle to defray operation and maintenance costs". In "special cases", where this is not possible, the Tanzanian government should provide the funds. The operationalization of these principles is still under discussion.

worsened. This illustrates the turbulence which affects even short-term planning, and which is often dictated by factors unrelated to project activities or project capacity to utilize funds.

On the basis of the three-year plans, one-year plans are then prepared in which detailed technical and manpower planning is undertaken. These plans are also used to coordinate the technical work with the work of the village participation coordinators. The plans are approved by the Regional Steering Committees established in each region (see below).

8.3 Institutional Framework

A number of agencies are involved – directly or indirectly - in the implementation of the DANIDA-funded project.[7] They include the existing Tanzanian organizations from village to central level (see Figure 3a), the donor organization itself, and five bodies established specifically for the water project:

– the Steering Unit (SU) placed in Dar es Salaam and staffed by DANIDA. The unit is responsible for project management, import of materials, and their distribution to the regions.

– the Implementation Office (IO), attached to the RWE's office in each region and staffed by engineers and mechanics from CCKK; a Village Participation Coordinators (VPC) and Tanzanian staff hired by DANIDA; and staff seconded from MAJI. The IO is the operative agent of DANIDA with regard to project implementation (planning, design, supervision, village participation, operation and maintenance of completed schemes, etc.).

– the Regional Steering Committee Meeting (RSCM), one in each region, with members from SU, IO, the Ministry and the regional authorities. It approves implementation plans and monitors project progress.

– the Village Water Committee (VWC) consisting of three men and three women from each village receiving a new water scheme. It

---

[7] Organizations related to health and sanitation are not dealt with here. Such activities have been rather limited, but will grow in importance from 1986.

represents the village vis-à-vis the authorities in matters relating to project activities.

- the annual joint Tanzanian/Danish Review Mission, consisting of DANIDA consultants and representatives from the Ministry. It reviews project activities and plays a significant role in making major management decisions.

The SU, the RSCM and the Review Mission were established during the preparation of the RWMPs. They were formed on the basis of the project appraisal (1979) and experiences with implementation in phase 1. The IO was established at the start of phase 2, based on proposals by the SU. Only the establishment of VWCs was proposed in the RWMPs.

The emergence of this structure illustrates three typical problems in donor aid to rural development: (i) the limited influence of long-range planning on organizational structures; (ii) the bypassing of existing agencies of both the donor and recipient; and (iii) the difficulties of finding an appropriate organizational placement of a donor project.

The RWMPs proposals have had a limited impact on organizational structures apart from the establishment of VWCs. Four other major organizational proposals in the RWMPs were not implemented. One proposal was to decentralize construction of small water schemes and the maintenance of most schemes to the district level. A second proposal was to concentrate implementation during phase 2 in one "experimental" region at first, in which a 7- to 10-member technical assistance team should assist the regional and district authorities to improve their implementation capacity. The experience and procedures developed here should then be transferred to the other regions after 2 to 3 years. The third proposal was to separate the water-related activities in the central ministry from other activities in the ministry (such as energy, minerals, etc.) by creating a Directorate for water activities (CCKK, 1982d, Ch. 10.3; Ch. 11.7.; Ch. 11.5.3). The fourth proposal was to make MAENDELEO responsible for village participation activities (IRA/CDR, Vol. 13, Ch. 4.).

Although all these proposals were formally approved as part of the RWMPs by the two governments, attempts to carry them out during phase 2 were not made. One reason is undoubtedly that the phase 2 rate of implementation has been considerably smaller than assumed in the RWMP (based on the 1991 goal). A more important reason appears to be the considerable inertia of existing organizational struc-

Year	Implementation (mill. D.kr.)	Expatriate professional staff	DANIDA-hired Tanzanian staff
1982	10−15	4 1/2	8
1984	31	12	100
1985	57	12	410

Sources: DANIDA Steering Unit (various reports)

tures, which – although perhaps not optimal – may be preferred to new and untried ones.

The second problem – organizational bypassing – is also illustrated by the Danish case. It even shows two versions of this problem. The creation of a donor-staffed, Dar es Salaam based, Steering Unit in 1980 to manage the project constituted a partial bypassing of the Tanzania-based DANIDA Mission. Although the latter has always had the formal donor responsibility for the project, the actual responsibility has been shared between the Mission, DANIDA headquarters and SU in ways that have sometimes been confusing and unclear to both insiders and outsiders. Several heads of mission have tried to solve this problem with varying degrees of success. By the end of 1985 the problem remained unsolved. However, despite such conflicts it is clear that the SU has contributed significantly to the project – both during phases 1 and 2. Not only has it been responsible for the importing and distribution of materials for the activities in the three regions, but it has also been instrumental as a coordinating link between the consultants and the Tanzanian authorities. Although there has been a tendency towards a technical bias the SU has performed an important guiding and supervisory role vis-à-vis the engineers and social scientists.[8] No other donor in the water sector has had a similar organizational arrangement.

It is perhaps surprising that the bypass problem which the SU created within DANIDA has not alerted agency staff to a similar bypass problem between the DANIDA-funded project and the Tanzanian authorities. The latter have namely increasingly been bypassed since 1980. From 1980 to 1985 the number of directly DANIDA-employed project staff has increased much more than the funds available for implementation.[9] By the end of 1985 a total of 12 expatriates

---
[8] The background of the expatriate staff in the SU is engineering and administration.
[9] The trend is shown in the table.

and 410 Tanzanians were directly employed by DANIDA. Some of them worked with SU in Dar es Salaam and others with the implementation offices in each of the three regions. The SU has increasingly bypassed the central level import organization (MAJI central stores) which is reported to be notoriously corrupt and inefficient. At the regional level the IOs have gradually increased their own capacity to plan and implement. In addition they have gradually taken greater control over some of the RWE staff. On the one hand this trend towards greater control reflects the deteriorating economic conditions in Tanzania. On the other hand, it implies that neither DANIDA nor the Tanzanian authorities have been willing to adjust the implementation rate to the capacity of RWEs and the central level agencies.

The bypassing contributes to the production efficiency for which the project has been praised (DANIDA, 1985b, 5). It also contributes to diminish misuse of project resources. However, bypassing does little to improve the capacity of central level institutions or the RWEs to secure material supply to construct new schemes, and to maintain them in the future. This, together with a lack of any systematic training activities ever since 1980, illustrates the problems of maintaining a proper balance between short-term project efficiency and long-term institutional development and project sustainability.

Finally, the Danish project illustrates some organizational placement problems that are typical of donor assistance to rural development – in particular the difficulties in working through decentralized and multi-sectoral structures. Thus, the Danish-funded project has been designated as a "national" project except for a one-year period in 1983/84 when it was a "regional" project (see Ch. 2). As a regional project the regional and district authorities under the Prime Minister's Office became the counterpart project agency. This created substantial administrative problems especially concerning transfers of funds from donors through the Treasury to the regions. It also meant that donor funds for the water sector had to be budgeted as ordinary regional development funds. In an attempt to reach an interregional balance in development expenditures, these must not exceed a total regional investment ceiling stipulated by the PMO. The status as a "national" project was therefore soon restored. To the regional authorities in Iringa, Mbeya and Ruvuma it meant that the ceilings on regional development expenditures could be circumvented. To the central Ministry of Water it meant a larger authority over project activities vis-à-vis the regional administrations. And to the donor it meant faster and safer administrative procedures. But the change from a regional to a national project status also contributed to separate the project activities from the local administrative and political context.

DANIDA has also preferred to work at and with the regional level (e.g. RWEs) rather than the district level (e.g. DWEs). This is contrary to the recommendation of the RWMP which propose that a substantial part of construction activities (all "small" schemes) and most maintenance activities be carried out at district level (CCKK, 1982d, Ch. 11.5.). However, during phase 2 the argument has been that it would be too expensive to place major activities at district level, given the limited technical and administrative capacity here. It would also be more difficult for DANIDA to control them. By the end of 1985 the DWE offices had not yet been involved in project activities to any significant degree.

The sectoral placement of project activities have likewise created problems. The RWMPs, for example, proposed that an extension service be created to facilitate contact between the villages and the project authorities and to assist villages to participate in project activities (IRA/CDR, 1983, Ch. 4). There are two options for the placement of this service in the existing structure: (i) as a part of the community development department in the RDD or DDD office; or (ii) as a special unit within the RWE office. The first option would require a close cooperation between the technical staff of MAJI and the community development staff of MAENDELEO. The second option would require the creation of a new staff category in MAJI because such extension activities have not hitherto been carried out in this strictly technical organization.

In the RWMPs the first option was recommended (IRA/CDR, 1983, Ch. 4). However, during phase 2 a temporary arrangement was chosen, similar to the one used in phase 1. Extension functions are carried out by Village Participation Coordinators and their temporarily-hired assistants. They are not a part of MAJI, nor the Community Development Department – but are attached to the donor-established Implementation offices in the three regions and to the SU in Dar es Salaam. Despite ad-hoc contacts with the Community Development Department no deliberate attempts at strengthening and increasing the crucial extension service capacity within the existing institutional framework have yet been undertaken. The donor has preferred a solution which is undoubtedly more efficient and more easily controlled in the short run, but which is only viable in the long run if a systematic institutional development effort is made.

8.4 Scheme Technology

The choice of technology was made by a joint Tanzanian-DANIDA appraisal mission in 1978 prior to the preparation of the RWMPs. This mission recommended the construction of low-cost gravity schemes and – where such schemes could not be built – the use of hand-pumped wells (DANIDA, 1978). The RWMP proposals are based on these recommendations.

During implementation the RWMP proposals have largely been followed. However, experience has shown that the potential for hand-pumped wells is significantly larger than anticipated by the RWMP planners. Consequently, the use of hand-pumped wells is now mainly based on trial-and-error in villages where the geology appears promising. In this respect implementation experience has proven to be a better guide than the village specific RWMP proposals.

8.5. Service Level

In the RWMPs the service level is basically determined by a distance criteria (400 metres to a water point) and a supply criteria (25 litres per person per day). During implementation both criteria have been adhered to. On this point, which is specifically mentioned in the government-to-government agreement, the RWMP is followed.

Whereas the 400 metres criterion is a standard one used throughout Tanzania, the 25 litres criterion is only used on the DANIDA-funded schemes. The Tanzanian standard is 30 litres. The use of the 25 litres criterion has caused some problems especially during 1984. They are illustrative of the problematic RWMP approval procedures in the Ministry of Water.

The RWMP planners arrived at the 25 litres criterion on the basis of an extensive field survey of household water consumption patterns (BRALUP/CDR,1982, Ch. 8). It clearly showed that the 30 litres criterion could safely be reduced. Similar surveys from East Africa indicated the same (White et al., 1972). During the RWMP preparations, these results were presented at several meetings with the Ministry. Its representatives objected to the proposed reductions. Nevertheless, the consultants wrote the reductions into the RWMPs because they were based on extensive surveys and would result in cost savings. The RWMPs were subsequently approved in toto by the two governments. However, the Tanzanian approval was made by the Project Preparation Division in the Ministry under which the formal responsibility for RWMP preparation is placed (see Ch. 2). During phase 2, when the plans for new schemes designed on basis of the 25 litres criterion were forwarded to the Construction Division in the same Ministry for approval, they were rejected. Delays resulted, until donor insistance on adherence to the 25 litres criterion prevailed.

8.6 Priorities
The RWMPs stipulate that the selection of villages for implementation should be based on three main factors: (i) need as indicated by the access to and quantity of water presently available and by the occurrence of certain water-related diseases; (ii) technology and cost, in that high cost and/or inappropriate technology (diesel pumps, etc.) will delay implementation; and (iii) village acceptance: only villages willing to sign a contract specifying village contributions to construction, operation and maintenance will be eligible for a new scheme. The data on which the priorities were made were collected before 1982 and have not been updated since then.

The priority criteria are referred to in the government-to-government agreement and they have largely been followed during implementation. Undoubtedly the main reason is that there is no ranking among the priority villages. This provides room for considerable freedom in selection.

8.7 Construction
The number of people supplied with water from DANIDA-funded schemes is indicated in Table 10. As discussed in Chapter 8.3 the organizational framework for construction has not been based on the RWMP recommendations. During phase 2 the IOs have increasingly acted as executive agencies within the RWE organization, using a substantial part of the RWE staff and facilities. In this way the role of the RWEs has gradually been limited to provide skilled and unskilled manpower for field activities. This bypassing and displacement of the RWE was also discussed in Chapter 8.3.

There is one major exception to this bypassing trend. The construction of hand-pumped wells is carried out within the RWE organization with little direct involvement by the IO. The SU does, however, act as the material and equipment supplier. Yet without much technical assistance the RWEs in the three regions have constructed hand-pumped wells for approximately one-third of the 140,000 people being supplied by September 1985. It illustrates that some parts of the RWE organizations are able to produce results if provided with adequate back up.[10]

It is not difficult to identify the main factors which tend to force the project to bypass local institutions. First of all the two governments have not provided the project with specific development-oriented objectives. Without them the project management, encouraged by the annual reviews by joint Tanzanian-Danish missions, have emphasized production of new schemes. Furthermore, the implementation rate in phase 2 was only to some extent determined by the implementation capacity of the RWEs. Thus, at the end of phase 1 the technical advisors in the SU concluded that the RWEs in the three regions together had the capacity to build schemes in around 20 villages every year. This is one third of the implementation rate implied by the government-to-government agreement of 1983 (300 villages in five years). Obviously, technical assistance staff was needed to compensate for the capacity gap. Moreover, experience soon proved that the IOs needed a significant degree of independence from the local administrative structures and procedures in order to keep targets and to provide the desired level of control.

In this context it is noteworthy that so far the DWE offices have been even less involved in project activities than the RWE offices, although this varies from region to region.[11] At both levels very little systematic training of staff takes place. On-the-job training – to the extent that it is explicitly done – has been limited to the counterpart engineers, and sporadic training workshops for vehicle mechanics in some regional yards.

8.8 Operation and Maintenance

The poor state of many village water supplies all over the three regions in the early 1980s indicated the severity of the O & M problem (CCKK, 1982c, Ch. 6.3). To the RWMP planners it was clear that the Tanzanian approach to O & M was in need of major changes. They

---

[10] Some DANIDA staff have been concerned about the location of shallow wells within villages resulting in low service levels – a problem found in other shallow well projects, too (see Chapters 4 and 5).
[11] In Iringa, the DWEs are not involved at all. In Ruvuma they are involved to a limited extent. In Mbeya the DWEs are often directly involved in both the planning and construction of DANIDA funded projects in their districts. This development has not been pre-planned. It has grown out of the specific circumstances and experiences in the three regions.

agreed on three major principles: (i) the village – not the Ministry, the region or the district authorities – should be the owner of the water supply scheme and therefore responsible for operation and maintenance; (ii) villages should contribute in cash or kind to O & M; and (iii) the district – not the regional – MAJI organization should assist villages if these were willing to pay for O & M services.

Both the SU and the IO have tried to establish an O & M system in phase 2, but the translation of the RWMP principles into practice has caused considerable problems. On the one hand, knowledge about the effects and applicability of the three principles in the Tanzanian setting are very limited. There are no past Tanzanian experiences based on such principles to analyse. Even analyses of the existing O & M system are hampered by lack of data and by the complexity of the technical, economic, organizational and political issues involved.[12] On the other hand, political approval of the proposed principles are needed because they constitute a clear break with the previous Tanzanian policy of water as a free public service.

Not surprisingly, therefore, very little actual field experience with the village-based O & M system had been gained by the end of 1985, despite five years' implementation and a substantial effort at planning during phase 1. The O & M problems remained unsolved. Only a few schemes had been handed over to the villages and only one mobile unit had been established.[13] It is too early to judge how this system will work. The slow development of the O & M component is partly a result of the high priority given to construction and rehabilitation activities. It has kept the SU and the Implementation Offices fully occupied. But it is also a result of the difficulties on Tanzania's part in providing the project – and other donors – with more explicit guidelines as to how a village-based O & M system should be organized. It has obviously added to the uncertainties that a major change in the political and administrative structure was introduced with the reestablishment of Local Governments in 1984. Their role in rural water activities has yet to be clearly specified although the district council will be the focal agency in the new O&M set-up, according to the local Government Act.

[12] Except for socio-economic analysis and the proposals arising from them (BRALUP/CDR, 1982; IRA/CDR, 1983), very limited O & M analyses were made in the RWMP (see CCKK, 1982d).
[13] DANIDA has taken a strict project view on its support to O & M. It will establish mobile maintenance units to service mainly DANIDA-funded schemes. They will be operated by the IOs and will carry out preventive maintenance on a regular basis and corrective maintenance on village request.

8.9. Participation

The Danish approach to participation and its planning differs from those of other donors in the water sector in two ways. The development of the approach was based on direct field experience in which the socio-economic planners took an active part in the construction of new schemes. The normal separation of planning and implementation in the preparation of RWMPs was not maintained. Secondly, the approach was made operational by specifying the organizational structure needed to implement it and by providing a detailed procedure to be followed by the involved agencies in the planning, construction and maintenance of schemes.

During phase 2 many of the proposals on participation contained in the RWMP have been implemented (IRA/CDR, Vol. 13, Ch. 3) with the active support of the SU and the Review Mission. With respect to structural recommendations the proposed extension service function has been established (see Ch. 8.3). But being staffed by DANIDA-employed Village Participation Coordinators and assistants attached to the Implementation Offices at the regional level, it is a temporary unit similar to the one that the socio-economic planners operated during phase 1. It bypasses MAENDELEO, as discussed in greater detail in Chapter 8.3.

The other major structural recommendation in the RWMP related to participation concerns the establishment of Village Water Committees. A committee should consist of three women and three men and should specifically deal with water supplies and related issues. It should represent the village vis-à-vis the authorities and should be responsible for organizing the activities which should be undertaken by the village during the planning, construction and maintenance of a water scheme. This proposal was based on three major assumptions. The special VWC would be more active in dealing with village responsibilities than the existing village committees that often appeared dormant. Second, by including females in the VWC, the women would gain influence on water supply matters. Existing committees were almost exclusively male, and women's influence on village affairs therefore limited. Finally it was assumed that during the planning and construction of schemes the VWC would work closely with the extension staff and thereby gain the skills and authority needed to manage village responsibilities for O & M once the construction crew had left and the scheme was handed over to the village.

In 1982, when the VWC proposal was made, the socio-economic group only had experience with village involvement in planning and construction in a few villages. This experience was fairly positive. By the end of 1985 it was clear that although a VWC is an essential part of the participation approach, it is no panacea. VWCs vary widely in their skills, authority and motivation. Often the women play only a passive role in the VWCs – sometimes none. And many VWCs may slowly cease to function if left to themselves. It was becoming clear that VWCs cannot be expected to function without continued support from the outside. In other words, a deliberate effort at institutionalizing the VWCs is needed. It will not only require regular visits by the extension staff, more training of VWC members, and perhaps more extensive efforts at involving women at the sub-village level. It will also require a systematic attempt at reorientating the MAJI and CCKK staff towards a participatory approach. So far the village participation activities have tended to be an additional project component – not an integrated part of all activities.

With respect to procedures, most of the recommendations in the RWMPs and contained in the project handbook on participation are used by the Village Participation Coordinators in their day-to-day work. But there is one noteworthy exception. In the RWMP it is proposed that only villages willing to make a cash deposit equivalent to one year's O & M cost *prior* to the construction are eligible as candidates for new schemes. This, the socio-economic planners assumed, was the only way in which to test the future willingness and ability of villages to pay for the maintenance of their schemes. Instead of this fairly radical proposal it was decided that the willingness of villagers to sign an agreement stipulating village O & M responsibility, combined with willingness to contribute labour during construction, was sufficient evidence of village ability and willingness to pay for future O & M expenses. Since only very few schemes have yet been handed over to villagers (Ch. 8.8) this evidence is yet to be substantiated.

8.10. Coordination

Lack of coordination is typically a serious problem in rural development projects. But with respect to inter-agency coordination the RWMPs are almost completely silent. One major proposal by CCKK (1982d, Ch. 11.5.6) was to create a zonal coordinating structure for the three regions involved. It has never been implemented. The other major proposal, made by IRA/CDR (1983), – on the coordination of MAJI, MAENDELEO and AFYA activities at the village level – has not been implemented either, as already discussed. Instead the coordination problems and the attempts to solve them have grown out of the implementation experience itself.

As mentioned earlier, the Regional Steering Committees (RSCMs) were established to provide coordination already during phase 1. According to the government-to-government agreement the RSCM should continue during phase 2. It shall approve the three- and one-year implementation plans and monitor their implementation. By establishing the RSCMs the two governments have, however, bypassed the existing decision-making and monitoring system at regional and district level (see Figure 3b). It is not possible to judge what effects this bypassing has had. On the one hand the existing regional and district committees would no doubt be difficult and cumbersome to work through. It is also uncertain how active they would be in dealing with project decisions. On the other hand, the RSCMs have not been very active in either decision-making, monitoring or coordination. They have mainly had a legitimizing role, approving activities already planned by the IOs and the SU.

This does not appear to concern Tanzanian authorities overly much. Regional and district politicians and administrators tend to express appreciation of the progress in constructing new schemes and only a few express the need to influence what most regard as "DANIDA's project".

8.11. Monitoring and Evaluation

Neither the RWMPs[14] nor the three-year and one-year implementation plans provide any specific proposals on monitoring and evaluation. The current monitoring and evaluation activities have therefore grown out of experience from phase 1 and 2.

Regular reporting on project activities are made by the VPC, the IO, the RWE and the SU. The format and circulation of their reports change occasionally but two trends are emerging. One is that the directly-involved parties (IO, VPC, and RWEs) are slowly reporting in a more unified manner and may eventually produce common reports. The other trend is that reports have a wider circulation. The RSCMs specifically are now receiving reports from the IOs that were formerly only forwarded to the SU.

As might be expected from earlier chapters, the reporting is strongly focused on construction and its problems. This bias is best illustrated

[14] Proposals for monitoring and evaluation in IRA/CDR (1983, Ch. 13) have not been implemented.

by the lack of any regular reporting on the functioning of already existing schemes (more than 140,000 people were assumed to be supplied by the end of 1985).

Another characteristic of the reporting is its weak relation to the various plans – the RWMPs, the three- and one-year implementation plans. Partly this reflects the lack of specific and operationalized objectives in these plans. But it also reflects the limited use of the plans as tools for project management and implementation. Thus the reporting does not specifically inform about what was planned; what was actually achieved compared to the plans; the reasons for divergences; and the actions taken as a result.

While this lack of problem- and plan-oriented reporting is a general one, it should be added that on the key issue of financial reporting, this type of management-oriented monitoring has not been possible. The problem starts with the RWMP. Although this plan does indicate the construction cost of each village water scheme, the way in which these estimates have been prepared makes comparisons between planned and actual costs difficult. In addition, both MAJI's and DANIDA's accounting systems have been too aggregated to allow a scheme-by-scheme accounting of costs. As a result, neither the total costs of all construction activities nor the total construction costs of individual schemes have been calculated and compared to the RWMP estimates on a regular basis. This has substantially weakened the possibility of both the donor and the recipient to exercise the important budget control function. This is, however, a general problem in Danish project aid (Rigsrevisionen, 1983, 118).

Efforts are now being made by DANIDA to introduce a more suitable accounting system. In addition, an Annual Joint Tanzanian/Danish Review Mission has been monitoring the project. Gradually this mission has become an important forum for decision-making in the project and this has reduced the role of the SU and the RSCMs correspondingly. The ascendancy of the Review Mission, like many other important developments in the project since 1980, has grown out of past experience and is not a result of deliberate planning.

8.12. Key Observations

The Danish-funded RWMPs and the subsequent implementation in the three regions score fairly high marks among those Tanzanian civil servants who have first-hand experience with similar donor-funded activities in other regions.

It is true that without the RWMPs exercise it is unlikely that a Danish commitment on 250 mill. D.kr. would have been made for phase 2. Analyses in this chapter also confirm that the Danish-funded water project uses the RWMPs to a larger extent than is the case with other RWMPs investigated. This is particularly true with respect to priority and design criteria and to procedures for participation. The project's usage of the remaining parts of the RWMPs is fairly limited. Furthermore, the RWMP is used by DANIDA only. Other donors, and the local authorities implementing locally-funded rural water supply activities, do not appear to use the RWMPs to any significant degree.

It is quite clear that the amount of data collected and analysed during the RWMP exercise, outstrips actual planning and implementation needs. Only a limited set of hydrologic data are now updated on a regular basis in each region. Otherwise, updating of water-resource and socio-economic data is strictly geared to actual implementation. Most data not regularly updated will rapidly become obsolete, but proponents of RWMPs will claim that such data may prove, nevertheless, useful in the future.

In contrast to the massive data collection efforts made, the RWMPs contain few specific proposals on key issues like training and institutional development. Likewise, the engineering part of the RWMPs contain only few and general O&M recommendations.

One trend in the development of the project stands out very clearly. It is implementation experiences from phase 1 and RWMP proposals based on these experiences that have most directly influenced phase 2 activities. In contrast, RWMP proposals based on analysis only or proposals untried in practice have had little impact.

It is also clear that the project now develops without clear operational long-term guidelines. In their absence the emphasis has increasingly been to construct a substantial number of new schemes within the funds and time allocated. Long-term guidelines are needed to strike a balance between production of new schemes and institutional development efforts; between donor control over implementation and the eventual need to integrate project activities in the recipient organizations on the national level and downwards; and between a donor-financed welfare approach to service provision and an approach that assists self-help. Such development-oriented long-term guidelines are lacking from Tanzania's sector policies; in the RWMPs; in the government-to-government agreement; and in the three-year rolling implementation plans produced so far.

Two fundamental lessons can be drawn from the Danish-funded project. Planning based on implementation experience rather than extensive pre-implementation analysis is crucial for project development. Moreover, long-term guidelines that address major sector policy issues rather than the minutiae of specific project activities are needed to direct short-term decision-making. These issues are discussed in more detail in Chapters 11 and 12.

Part Three
Interpretations

9. Achievements and Failures of Control-Oriented Planning and Implementation

It is now possible to make some interpretations of the various planning and implementation approaches studied in Chapters 4 to 8.

9.1. Classification of Case Studies

The classification is based on the five dimensions of various planning and implementation approaches presented in Chapter 3.

Most of the medium- and long-term plans prepared between 1972 and 1983 are strictly functional. There are only a few examples of means-ends analyses of major sector objectives (such as water for all by 1991; water as a free public service, etc.). These exceptions are the plans funded by the World Bank (Ch. 7.8) and Denmark (Ch. 8.8−9). All other planning teams accepted the Tanzanian sector policies unquestioningly. This changed around 1983 when all remaining donors included in this study began to scrutinize sector policies; press for changes in them; and adjust their own assistance approaches (e.g. Finland in 1983–1984; Holland in 1984; and Sweden in 1982–1984).

A clear – but not complete – blueprint orientation has dominated all the five donor-funded medium- and long-term planning efforts. This is reflected in the detailed specifications of future production targets (e.g. number of new schemes to be constructed per year). In contrast, the inputs to and outcomes of other essential future activities such as training, participation, organizational procedures, have normally not been specified in similar detail (although see Ch. 7 and 8.9).

Another key feature of the blueprint orientation is the separation of planning and implementation activities both organizationally and over time. Every medium- and long-term plan was prepared by technical assistance staff working fairly independently of the recipient organizations. Moreover, every plan was made prior to any implementation – except the Danish preparation of participation procedures.[1]

It should be added that during implementation no continuous monitoring and formative evaluation of field activities and their outcomes – such as the functioning and utilization of completed water schemes – have been undertaken in any of the cases studied. Moreover, with one exception, no regular updating of the RWMPs or other medium- and long-term plans have been made. The exception is the RWMPs of 1976 for Mtwara-Lindi which are now (1985) being updated by the Finns.

None of the medium- and long-term plans are based on rational-comprehensive analysis. Some of the planning teams attempted such comprehensiveness on selected issues notably related to technical matters (water resources, scheme technology, etc.). However, none of the plans analyse and present major alternative means to reach given ends.

The non-participatory approach to planning and implementation has prevailed until 1983 in three of the five cases studied. Some degree of beneficiary involvement has, however, been strived at in the project planned and implemented by the World Bank[2] and the Danes. From 1983 onwards attempts have been made in all the cases studied to mobilize intended beneficiaries to take part in selected project activities.

Bypassing of recipient organizations is a key feature in all cases studied. Without exception this approach has been used during the preparation of all the medium- and long-term plans (and in the preparation of most one-year plans as well). This planning activity has invariably been controlled by technical assistance staff with a limited number of Tanzanian counterparts playing minor roles only. A similar bypassing approach has been used during the subsequent implementation. Donor-hired technical assistance staff has had complete control over the Finnish- and Dutch-funded projects until 1985 – although this is now changing. A less pronounced but still clear bypassing has been practised by the World Bank and the Danes. A somewhat higher degree of integration into local institutions is now (since 1983) being tried by the Swedes.

---

[1] Danida funded some implementation of new schemes concommitantly with the funding of the RWMP activities (Ch. 8). However, only the socio-economic team was involved in both activities. The technical planners were not involved in implementation activities.

[2] In this project beneficiaries contribute a part of the cost of hand-pumps to qualify for wells.

Various degrees of control orientation therefore characterize the planning and implementation approaches chosen by the donors in the five case studies. With the simplification inherent in all such classifications this control-oriented approach is characterized by (i) donor attempts at detailed *pre*-planning of some – but not all – future activities; (ii) major donor control over medium- and long-term planning activities and their subsequent implementation through the establishment of project management units with some (sometimes considerable) independence from recipient institutions; (iii) focus on assisting specific projects through these units rather than support for programme activities carried out by local institutions; and (iv) either complete neglect of beneficiary participation or emphasis on participation closely controlled from above.

9.2. Achievements and Failures

Has this control-oriented planning and implementation carried out with donor assistance during the last decade had any significant usefulness? In the absence of explicit evaluation methodologies for planning and plans (Killick and Kinyua, 1980), two different approaches may be used in answering the question. One is to try to estimate what would have happened without the plans. This is clearly very difficult. The other is to base the assessment on a set of normative questions. Both approaches are used below.

First, it is likely that the extensive donor involvement in and control over medium- and long-term planning at regional level may have increased the external support to the sector above the level that would have resulted without such involvement. From a purely sectoral point of view this may be regarded as an achievement. It was certainly one of the results hoped for when the Ministry invited donors to get involved in this type of planning with only few restrictions in terms of guidelines and coordinating initiatives (see Ch. 2.2).[3] Table 2 indicates that donor aid to the rural water sector did rise sharply during the 1970s.[4] Donor involvement in medium- and long-term rural water sector planning may have contributed to this trend by directly increasing the

---

[3] Thus the Acting Director of Project Planning Division in the Ministry stated that "Most of the donors approached accepted to do regional water master plans in specific regions and later on to do implementation ... This pays politically. The donors contribute 80 per cent of the capital investments to the water sector" (Balaile, 1983, 41).

[4] Measured in real terms donor support declined in the early 1980s (see Table 2).

total donor aid to the sector, or by causing a reallocation to the rural water sector of aid earmarked to Tanzania. However, the rise may also have been the result of the combined effects of a general increase in aid to Tanzania and a certain shift in aid towards rural development activities in general. Available data do not allow any firm conclusions on this.

Second, the heavy emphasis by all donors on substantial data collection and analyses during plan preparation may be claimed to have some utility. Thus, the water resources data are – by some (see Lium and Skofteland, 1983) – regarded as potentially useful. It is difficult to access this claim. On the one hand, much of the water resources data collected cannot be used or is not needed in the actual planning of rural water supplies (see Ch. 10.3). On the other hand, time series data on water resources are useful if – in the future – major water resource demanding activities are to be planned (large scale industry, irrigation, etc.). In this respect the cost of collecting the data could be regarded as an insurance premium against unforeseeable future data needs. Certainly some collection of water resource data is justified, especially if it is done on a continuous basis. The inventories of existing water schemes collected during planning have also been useful. Finally, it is clear that some of the socio-economic data have been useful (Ch. 8). But the collection and analyses of socio-economic data are not uniquely tied to any particular planning and implementation approach. Rather it is the *volume* of data collected and analysed that can be related to particular planning and implementation approaches. And in this respect there is little doubt that the control-oriented approaches to planning as used in the cases have led to excessive data collection and analyses.[5] Interestingly, this emphasis on data has been much less pronounced during *implementation*. As already mentioned in Chapter 9.1, very little effort has been put into monitoring and formative evaluation.

Third, donor involvement may have helped to induce certain important changes in the rural water sector. Thus the Dutch have contributed significantly to the spread of the hand-pumped well technology in many regions. This technology has now also been accepted – although sometimes without much enthusiasm – by the Tanzanian authorities. The World Bank has introduced the concept of user payment for water supplies. Several other donors have subsequently

---

[5] Although in general the amount of data actually analysed during the planning stage is considerably less than the amount of data collected.

introduced their own versions of this concept in the regions they assist. The Party and the government have not yet agreed on a countrywide change in the free service policy although it is being intensively discussed. Finally, the Danes have operationalized the participation concept in rural water schemes. The extent of this participation and the experiences with it are still limited but the approach and procedures have been adopted in some other regions and have been accepted at the central level. But it should be emphasized that the two latter innovations have not yet resulted in any official changes in rural water sector policy. However, these three important changes are not necessarily the result of specific donor approaches to planning and implementation as such. They are rather the result of particular donor policies and of the substantial donor involvement in the rural water sector.[6]

The assessments of the various donor approaches to planning and implementation can also be based on normative criteria. They are selected from Belshaw (1979) and Cairncross *et al.* (1980) and are based on the officially-stated objectives of the Water Master Planning exercise, that the plans should "provide the Government of Tanzania with firm recommendations for the immediate and long-term development of water resources with particular reference to human and livestock use" (Balaile, 1983). Here the criteria are formulated as questions, as indicated in Table 11.

Questions (a) to (d) concern the formal properties of the plans. Three of the plans included in the case studies are functional and therefore consistent with national targets and policies, whereas two others are not (see a). It is not possible to ascertain to which extent the plans are consistent with regional objectives (see b); partly because such objectives are not clearly stated in an operational way by the regional authorities themselves, and partly because the rural water sector plans were prepared fairly independently of the existing regional planning framework. Even in cases where the medium- and long-term plans have been "approved" by the regional authorities this may not necessarily have ensured consistency (see discussion in Ch. 2.2–3). None of the plans are consistent across regions (see c) because each planning team has been working according to its own terms of reference and plan formats. Finally, it is noteworthy (question d) that only

---

[6] Swedish attempts to integrate water, health and sanitation activities is a fourth example of a donor-induced change. However, it is too early to assess whether the attempt will succeed and whether the integration approach will be taken up by other donors or by the government of Tanzania.

Table 11. *Summary Assessment of Control-Oriented Planning and Implementation Approaches*

Have approaches assisted in the identification and design of rural water sector strategies which are:[a]	FINNIDA[b] (Ch. 4)	DGIS[c] (Ch. 5)	SIDA[d] (Ch. 6)	World Bank[g] (Ch. 7)	DANIDA[h] (Ch. 8)	Comments
(a) consistent with national target and policies?	yes	yes	yes	no	partly	All RWMPs based on national target (Water to all by 1991)
(b) consistent with regional authorities' development objectives?	?	?	?	?	?	Some medium- and long-term plans approved at regional level; others not. Only the World Bank plan is explicitly linked to regional development plans or to Regional Integrated Development Plans (RIDEPs)
(c) consistent across regions?	no	no	no	no	no	No attempts at such coordination were made by the individual planning teams. Central level attempts at coordination during early 1980s failed
(d) based on specific objectives of intended beneficiaries?	Not considered in plan	Not considered in plan	Not considered in plan	Cash payment qualifies for new schemes	Left to implementation stage	
(e) appropriate to regional water resources?	partly	n.i.	n.i.	n.i.	partly	

158

					Comments
(f) appropriate to financial and institutional capacity at:					
national level?	no	no	n.a.	no	Production targets made independently of local capacity
regional level?	no	no	no	no	All plans are biased towards production
(g) effective in identifying conflicts and trade-offs between objectives and constraints?	no	no	no	no	
(h) Has donor-assisted implementation been based on medium and long term plans with respect to:					
– institutional development	no	no	no	no	
– technology?	partly	yes	no[c]/yes[f]	partly	Present Swedish implementation not based on initial RWMP
– service level?	partly	partly	n.a.	yes	
– priorities?	unclear	unclear	n.a.	yes	
– construction?	partly	yes	n.a.	no	
– operation and maintenance?	no proposal	no proposal	n.a.	yes	
– participation?	no proposal	no proposal	n.a.	no	
– monitoring and evaluation?	no proposal	no proposal	yes	no	
(i) Have locally-financed implementation been based on the medium- and long-term plans prepared by donors for the respective regions:	no	no	no	no	
(j) Have donor planning and implementation approaches contributed to sustainability	n.a.	n.a.	no	n.a.	It is to early to judge the Danish/Swedish project on this point

n.i.: not investigated
n.a.: not applicable
a Partly based on Belshaw (1979, 10) and Cairncross *et al.* (1982, 2).
b Mtwara-Lindi RWMP
c Morogoro Domestic Water Supply Plan
d Lake regions RWMP and VIAK plan
e RWMP
f VIAK plan
g Water Supply Component Plan
h Iringa, Mbeya and Ruvuma RWMP

two of the plans provide a structure which allow the implementors to consider beneficiary interests – at least to a certain extent.

Questions (e) to (g) concern the type of analyses on which the plans are based. With respect to analyses of water resources (see e) there appear to be some problems in the two plans from which information is available. It is not possible during a relatively short planning period to collect and analyse sufficiently reliable and detailed microlevel data for the planning of individual rural water supply schemes (Ch. 4, 8 and Ch. 10.3). None of the plans have provided sufficient analyses of various financial and administrative constraints, and have therefore not been able to provide proposals on how trade-offs between constraints and objectives could be made (see f and g). Much of the planning exercise has consisted in working backwards from the 1991 target to the resources and capacities required to reach this goal regardless of the feasibility of the resulting demands on funds, manpower, equipment, etc.

Questions (h) and (i) concern the extent to which proposals on key issues have been implemented. With respect to the donor-funded implementation some limited plan compliance can be observed. However, available evidence suggests that these plans are used even less or not at all when implementation is funded out of non-donor funds.

The final question (j) concerns sustainability. The case studies clearly show that approaches used by Finland, Holland and the World Bank were not sustainable. The two former donors are now trying to introduce major changes in implementation approaches while the latter has pulled out completely. It is too early to judge if the Swedes and the Danes will do better with respect to the long-term sustainability of their assistance.

The picture which emerges from this assessment of the case studies is therefore a rather bleak one. The utility of the donor-funded medium- and long-term planning prepared prior to 1982 has been quite limited, as WHO/World Bank (1977), Balaile (1983) and Schoenborg (1983) have also concluded. The sustainability of the activities implemented through donor-controlled project units has also been a major problem throughout the 1970s and early 1980s. And finally, the quite substantial donor assistance to the rural water sector has not yet contributed significantly to a reformulation of the sector policies of 1971. In fact the donors have, until recently, supported and accepted the major elements of these policies such as the 1991 target and the concept of the free provision of water to beneficiaries. In this way their support has perhaps delayed major policy changes, the urgency of which has become increasingly clear through the 1970s (see Ch. 2).

It is this combination of an increasingly inappropriate sector policy combined with the particular donor-supported planning and implementation approaches described in the preceding chapters which have led to the present crisis in the rural water sector.

9.3. Persistence of Control Orientation

The major problems in the donor approaches to planning and implementation identified above are *not* specific to the rural water sector. Kleemeier (1984), for example, has reached similar conclusions in her research on donor involvement in regional integrated development projects in Tanzania. Moreover, research from other countries on medium- and long-range planning and implementation also indicates the problems of the limited usefulness of such plans and the lack of sustainability of donor-assisted activities. They were already identified by Waterston (1965) with respect to national- and sector-level economic plans, and his findings have been reconfirmed since then by, for example, Faber and Seers (1972); Caiden and Wildawsky (1974); Killick and Kinyua (1980); and Argawala (1983). Planning for rural development has generally not fared better, as testified by Chambers (1973), Lele (1975), Korten (1980), Johnston and Clark (1982) and Rondinelli (1983). This limited usefulness or outright non-use of plans and the problems with sustainability are therefore *not* specific to Tanzania, its rural water sector, or the donors and consultants engaged in it. Rather, it is the assumptions of the control-oriented planning and implementation approach which do not hold. This will be shown in the next chapter.

10. Limits of Control-Oriented Planning and Implementation

Many of the persistent problems of the control-oriented planning and implementation approach occur because of the uncertainty and complexity of the rural water sector activities illustrated in Chapter 2. Several of the basic assumptions of the control-oriented approach (Ch. 3) do not hold under these circumstances, as the analyses in this chapter will show.

10.1. Political and Bureaucratic Conflicts

The control-oriented planning and implementation approach is based on the assumption that a certain consensus on objectives exists (Ch. 3.1). Yet, even if such objectives are explicitly stated in policy documents from the donor and the recipient and written into commonly approved plans, they may often hide equally important – but unstated – objectives. The literature contains numerous examples of this (Morss and Gow, 1985). Much of the empirical evidence available from this study illustrates the problem, for differing agendas appear at all levels and in all phases of the planning and implementation activities. Here they are discussed from a recipient, a donor and a beneficiary perspective.

10.1.1. The Recipient
Tanzania has undoubtedly had several motives for inviting donors to prepare medium- and long-term rural water sector plans. Apart from the control function of such plans they may also serve political and bureaucratic ends. Friedmann (1967), for example, suggests ends such as: mobilization of external resources; increasing the influence of the agency (or agencies) involved in the planning; symbolic representation of progress; and helping to build national consensus.

The mobilizing effect of plans has already been discussed in Chapter 9.2. Evidence of the use of the medium- and long-term plans in bureaucratic struggles is quite clear. Thus the Ministry has gained a certain influence over rural water sector activities through the RWMP exercises. Designated as a "national project" (Ch. 2.3) they fall under the Ministry's auspices. In this way the central organization has maintained some of the control over regional activities that it has been losing to the regional authorities since the decentralization of 1972. Obviously the central level interest in planning as a control instrument does not necessarily fit the requirements at local level for operational plans. The case studies abound with examples of plan proposals not implemented at all.

The political purposes of planning in Tanzania were explicitly noted by Rweyemamu soon after independence. He focused on national planning but his observations appear equally valid in the context of rural water sector planning. In 1965 Rweyemamu wrote:[1]

> In a new nation like Tanzania, a national plan is a major, albeit incomplete, substitute for the goods which were promised explicitly or implicitly during the struggle for independence. In so far as it is indicative of a future of abundance, a national plan serves as an unifying agent of an otherwise loose and fragile society...

Hyden (1979) – also writing on Tanzania – adds, that by describing a better future, plans may serve to constrain the claims of various political interest groups wanting immediate satisfaction of their demand. The 1991 goal for the rural water sector; the policy of water as a free public service; and the Tanzanian insistence that medium- and long-term planning be based on these elements, can best be understood from this perspective. It makes the State the benevolent provider of services to beneficiaries. This role is perhaps central to the understanding of the State in Tanzania (Stein, 1985). But this ideology makes participatory planning and implementation difficult, although participation is also an important part of the official policy (see Ch. 10.4). Furthermore, it tends to divorce medium- and long-term plans from reality. This contributes to the non-use of plans as illustrated in several of the case studies (see Chapters 4, 5, 6 and 8).

The impossibility of reaching the 1991 goal and providing people with "free" services has been known to the Ministry since the mid-1970s. At an internal meeting between the Ministry and all RWEs held to discuss President Nyerere's speech "The Arusha Declaration: Ten years after", it was concluded that:

---

[1] As quoted in Friedmann (1967).

either the necessary resources be provided on target or the target be restated. (Ministry of Water, Energy and Minerals, 1977)

This did not happen. Indeed it has not yet happened. The official line has been maintained in negotiations with donors despite clear awareness in the bureaucracy that the 1971 policies were becoming increasingly irrelevant (see for instance Msimbira, 1984). Thus three RWMPs are based on these sector policies (Ch. 4 to 6) while one RWMP (Ch. 8) is partly based on them. Yet the 1991 goal and other elements of the rural water sector policy are clearly political statements which are inappropriate for operational planning purposes. Only recently have the policies been challenged by donors.

Obviously the 1971 policies encourage a bias towards construction of new schemes rather than rehabilitating and maintaining existing ones. This bias is clear at national, regional and district levels. For example, even in cases where maintenance responsibilities for donor-funded activities have clearly been assigned to local authorities and written into donor-recipient agreements, Tanzanian authorities have ignored them. The Dutch case study clearly shows this (Ch. 5). Unfortunately, donors also tend to be biased towards constructing new schemes (see below), and this amplifies the problem. Furthermore, Table 2 indicates how recurrent expenditure allocations to rural water supplies have not kept up with inflation and the increase in the number of new schemes. Although such biases do not prevent planners from making medium- and long-term plans based on a proper balance between capital and recurrent expenditures, they certainly make it difficult to implement them. The World Bank experience in Mwanza (Ch. 7) illustrates the point.

Also more specific political and economic incentives amplify the *de facto* bias towards the construction of new schemes. Thus the Bank of Tanzania and the Ministry of Finance are concerned about total aggregate aid flows and the extent to which they may cover projected balance of payment and budget deficits (Jennings, 1981). Moreover, senior officials in the Ministry of Water seek specifically to maximize aid to the water sector. Construction of new schemes requires more funds (which can be spent faster) than does the maintenance and rehabilitation of existing schemes. Such considerations tend to lead the top decision makers to focus on the overall flow of aid. The question of how to utilize the funds most appropriately therefore does not receive the attention it deserves. Donors often share this interest in moving money (see below). This has obvious consequences for planning and implementation.

10.1.2. The Donors

Aid to rural water projects is, no doubt, based on a genuine donor interest in "improving the health of the people and . . . creating better prospects for social development and economic growth" as is, for example, the declared main goal of Swedish assistance to the water sector (SIDA; 1984, 1).

In many ways aid to this sector is rather attractive from a donor point of view. It is a visible sign of donor interest in rural development. A water supply is often high on the priority list of rural beneficiaries (BRALUP/CDR, 1982). Assistance to such supplies is therefore also fairly non-controversial from the point of view of both local and foreign elites. It does not challenge vested interests (Chambers, 1983, 164). Water schemes even benefit women. Finally, the International Drinking Water Supply and Sanitation Decade (Ch. 2.5) has helped to focus attention on the sector and the donors who support it.

But donors have additional items on the agenda which influence the way they plan and implement. Many recent studies of donor bureaucracies demonstrate that their incentive structures tend to reward the moving of money (Honadle and Klauss, 1979; Gray and Martens, 1983). It is commonly agreed, as Morgan (1983) states, that "project development is an advocacy process as well as one of critical appraisal". Seffin (1979)[2] laments that this tends to force planners to be assertive:

> *these* are the goals; *these* are the proper purposes; they *will* be served by these outputs; and the outputs *will* be produced by these inputs within the time-frame and this particular setting.

Morss (1984, 467) adds that the advocacy process also encourages a high degree of official optimism. Donor staff, he claims:

> could not get their projects through their own agency's approval process if they made realistic estimates of the recipient countries' abilities to make resource commitments to projects.

[2] As quoted in Morss and Gow (1985).

To get aid projects approved in the donor bureaucracy and the relevant political bodies, planners must therefore express certainty. It not only mocks the uncertainty and complexity of conditions in developing countries (see Ch. 2 and 10.3), it also results in plans and project papers that frequently are not useful for implementers. All case studies provide examples of this.

It should be noted that consultancy companies and research institutes involved in planning and implementation in the rural water sector in Tanzania have strong incentives to deliver what the client (the donor) wants (see Ch. 2.2). The quest for moving money fast also fits in with their legitimate business interests. It simply increases the demand for their services. Recipient interest in increasing gross aid flows has already been mentioned. A mutual bias towards the construction of new schemes thus exists. As a result *the key problem* in the rural water sector – operation and maintenance – has tended to be neglected both in the medium- and long-term plans (Ch. 4 to 6 and 8) and during actual implementation (Ch. 4 to 8). Moreover, the incentive to move money is in conflict with professed aims at involving beneficiaries, as discussed in Ch. 10.4.

From a donor point of view assistance to maintenance activities has additional disadvantages. It would require long-term commitments in which both the magnitude and duration of assistance would be very difficult to predict in advance. And it would require broader-based assistance to programmes and institutions rather than control-oriented project support. All these are commitments which political bodies in donor countries are reluctant to underwrite. Hence the customary recipe: donors assist in capital expenditure activities, and recipients take over maintenance responsibilities. Unfortunately, as the case studies show, this recipe is also reflected in the donor-funded planning which is heavily biased against maintenance issues.

Donors have an interest in a certain return flow of aid money, too. Construction of new schemes offers a good opportunity for this, because it requires substantive inputs of materials, equipment and technical assistance. Tanzania's ability to deliver physical inputs has declined rapidly ever since the mid-70s as a result of the economic crisis and the squeeze on foreign exchange. This has prevented the full utilization of already installed local production capacity for cement, pipes, etc. Together with the high construction targets aimed at in the donor-funded water projects, it has necessitated a substantial amount of importing of materials. Some 60 per cent of total donor expenditure on rural water projects in Tanzania (including the cost of technical assistance) may therefore flow back to the donor countries themselves, according to two recent evaluations (see Table 2, footnote b).

A vital link between the rural water sector and the domestic industrial sector has therefore not emerged in Tanzania. Even local production of hand-pumps has not materialized, although this was written into the plans of two donor projects (Ch. 4 and 5). The donor preference for setting up implementing structures that bypass existing organizations has undoubtedly also been influenced by the desire to control imports. This control aspect has several other dimensions, as discussed in Chapter 10.5.

10.1.3. The Beneficiaries

In control-oriented planning and implementation it is often assumed that the project outputs fit beneficiary needs. This may or may not be the case. If it is not, the consequences of ignoring beneficiary needs can often be considerable. In such cases intended beneficiaries may accept the services provided, but not care about their maintenance. This has often happened in Tanzania (Ch. 2.1). They may also be reluctant or even refuse to participate in planned activities (Ch. 7). Or they may sabotage the services provided or just not use them (Mujwahuzi, 1978). So, although beneficiaries rarely have the power to affect project activities significantly at the planning stage, they do often have the power to refuse to take part during the implementation stage. This contributes to the non-use of plans made with little or no participation of the intended beneficiaries.

10.1.4. Goal Displacement

Setting goals is the cornerstone of all planning (Heaver, 1982). In control-oriented planning and implementation it is assumed that such objectives are reasonably clear, consistent and based on some degree of consensus – whether they are set through a functional or a normative mode of planning.

But medium- and long-term plans based on declared common donor-recipient objectives may not necessarily be used during subsequent implementation. For this implementation takes place in a context of conflicting interest between recipient institutions, donors, consultants and beneficiaries. This often results in:

> a simplification of the agency's operational goal structure, eliminating many of the non-measurable or long-run goals and focusing attention

exclusively on activities that will yield short-run measurable results. Immediate results have a political utility for the organization regardless of the contribution that activity makes to the long-run success of its programme. (Quick, 1980, 57)

Several of the case studies exemplify this tendency. The World Bank experience (Ch. 7) is a very clear example. Such goal displacements cannot be completely avoided. But the control-oriented planning and implementation approach is badly suited to a context in which "rural development objectives are often multiple, ill-defined, and subject to negotiated change." (Korten, 1980, 497).

10.2. Multiple Decision Makers

Control-oriented planning and implementation approaches are also based on the assumption that agencies exist which have the authority, capacity, and incentives to plan and implement such plans (Ch. 3.2). Unfortunately there is an almost total lack of discussion of the nature of the State in the vast literature on policy-making, planning and administration (Saul, 1972, 18; Hyden, 1983, 57). The assumption that a strong State exists in Tanzania is perhaps based more on its numerical expansion, on the strong rhetoric used by the Party, and on the ideological biases of the observers, than on empirical evidence. Most African countries are characterized by a personalized, fragmented power structure (Ekeh, 1975; Hyden 1983). Tanzania is no exception here (Saul, 1972; Iliffe, 1979). The State is not yet dominated by any class with roots in the economic structure which could control state activities. Its economic position is weak and has been eroded by the growing economic crisis during the 1970s and 1980s.

Moreover, the institutional framework within which rural water sector activities are planned and implemented, is rather complex. Figures 2 and 3 indicate that the number of potential decision-makers involved is substantial. But even in cases where all parties involved may agree on a certain activity, the large number of agencies involved results in delays and changes. This tendency is amplified when all types of inputs are in short supply as is the case in Tanzania. During implementation this brings a project or a programme under intense pressure from individuals and interest groups seeking access to the resources provided.

The presence of donors adds to the number of decision-makers. By the end of the 1970s a total of 16 donors financed 46 different projects in the water sector (Mushi, 1982). Each project tends to march to "the procedural tune, reimbursement schedule, and particular objectives of the respective donors" (Honadle and Rosengard, 1983a). Attempts at donor coordination have been feeble and without much success (Ch. 2.3). As shown in the case studies, a cacophony of donor-financed plans, technologies and policies have emerged: Plans that are different in scope, methodology and detail; technologies that are often incompatible (each donor tends to import their own materials, fittings and pump types); and policies pursued by the donors that are not only different from but sometimes contravene official Tanzanian policies. The World Bank, for example, demanded advance user payment for hand-pumps in their Mwanza project (Ch. 7), while villagers in Mtwara-Lindi were paid to excavate trenches by the Finnish project (Ch. 4). Thus the World Bank introduced direct user payment and the Finns abolished self-help. Both elements contradicted key aspects of Tanzanian rural development policies at the time.

In this context no authoritative decision-maker (such as a ruling class, Supreme Party, President, sectoral ministry) exists to set objectives, allocate resources accordingly, and to control the means to implement activities according to plans. Instead, rural water supply activities involve a staggering variety of local and external organizations, interest groups and people all pushing and pulling in pursuit of their various interests. In these circumstances conflicts over development activities typically emerge *during the implementation stage rather than during the planning stage.*

Control-oriented planning and implementation breaks down under such conditions. Planned activities typically undergo drastic alterations during implementation due to this pressure. Frequent adjustments become necessary and explain, perhaps, the observation by Moris (1977, 79) on recipient organizations:

> there is a flexible attitude towards plans, which are viewed as paper commitments mirroring a certain situation in power relations at one point in time.

The case studies provide several examples of the often drastic changes of medium- and long-term plans that occur during implementation (especially the case studies on Finland, the World Bank and Sweden). It adds to the turbulence when donor policies undergo substantial changes over time as well (see Ch. 2.4).

The existence of numerous decision-makers and continuous conflicts about development activities during implementation cannot be overlooked. They are key characteristics of the rural development context (Grindle, 1980, 15–19) to which planning and implementation approaches should be adjusted, as discussed in Chapter 11.

10.3. Uncertainty and Complexity

The preparation of control-oriented medium- and long-term plans requires predictability, knowledge and information about the various interrelationships between, and outcomes of, the activities being planned (Ch. 3.3). Such assumptions are rarely, if ever, met in the context of rural development.

10.3.1. Unpredictability

It is often argued that the non-use of plans and the non-sustainability of implemented activities are due to drastic changes in political, economic or administrative conditions that cannot possibly be foreseen. This is quite true. Planners working in Tanzania in the mid-1970s, for example, could not predict, say, the economic crisis of the late 1970s and its devastating effect on foreign exchange, imports, sparepart supply, work discipline and so on. They could, however, have predicted the unpredictability of future events. It is the assumption about stability which is the problem in medium- and long-term planning of rural development projects. This problem is reinforced by the lack of monitoring and evaluation during implementation (see below).

10.3.2. Poor knowledge

Lack of knowledge is another serious obstacle to control-oriented planning and implementation. The case studies provide several illustrative examples. Even within a supposedly well-established field like hydrogeology the experts had difficulties:

> Approximately 40 per cent of the wells installed in Shinyanga by the end of 1976 were not functioning properly . . . [This is partly] the result of errors in survey and siting, reflecting inadequate knowledge of the behaviour of local aquifers . . . (World Bank, 1977b, 3)

The Finnish team in Mtwara-Lindi experienced exactly the same type of problem (Ch. 4). In the Danish funded water project it is now realized that the knowledge about surface water resources is inadequate and that the information on this presented in the RWMP is not sufficiently reliable for the planning of small rural water schemes (Ch. 8).

Some planners also struggled hard to specify the interrelationships between water, agriculture and population growth and distribution. One team (NEDECO, 1974, 86) stated that one aim of the RWMP was "the optimization of the average income derived from agriculture". Consequently:

> The method applied is based on an optimum use of available lands, capital and labour, whichever is scarcest. Promising crops, yield projections, input requirements, availability of land, water, labour and capital, prices and market projections, and changes in the composition of the livestock herd are estimated.

The results of this model were then fed into a model of land-carrying capacity in order to estimate population growth and average incomes in various subareas of the region over a 20-year period. Even gross in- and outmigrations were estimated. The planners then recommended – based on such calculations – the yearly establishment of 20 new villages over a 20-year period until 1991. This, finally, became the basis for village water supply plans. One need not know much about Tanzania or population projection models to realize that a good deal of voodoo must have gone into the specification and quantification of the model.

The Dutch team was not alone in writing double-Dutch. The extract below is taken from the deliberations of the Swedish consultant when attempting to make a cost-benefit analysis of rural water supplies in the Lake regions:

> The drive to modernize that is taking place in rural Tanzania is first of all a change in the structure and tempo of life. Marginal economic calculations which assume little change are probably not very useful. This, however, leaves project [appraisal] rather up in the air. We believe that the entrepreneurship and imagination of the population is such that the value of saved time is greater than 30 [Tanzanian] cents per hour. (Brokonsult, 1978a, 5.10)

Lack of knowledge about water use patterns among villagers had more serious consequences. Many RWMPs and implementation plans proposed that one improved water source be installed in each village in the first stage of water supply development. This was in agreement with Tanzanian policies from 1975 to 1981. It was assumed that women would then use this one improved source instead of the traditional and often polluted ones. Just a few observations of actual water use patterns in villages would have shown that this assumption is simply wrong. Only rarely do women bypass a traditional source to draw water at an improved one (BRALUP/CDR, 1982, Ch. 4). Such lessons about the social and cultural aspects of water use were normally only learned well into the period of plan implementation.

One reason for this lack of knowledge is that Tanzania has not had professional staff in sufficient numbers to prepare the plans needed for rural water sector development (Ch. 2.3). An equally important reason is that the foreign technical assistance staff used instead, has often been rather inexperienced in preparing medium- and long-term rural water sector plans. This has been the case for much of the personnel from the Nordic countries (Wingaard, 1983). In general it should not be assumed that a formally well qualified "expert" in a particular field necessarily has adequate country- and locality-specific knowledge. It is a myth to believe that developed countries have a large pool of experts with a ready-to-use knowledge of rural development activities (Lethem and Cooper, 1983). The third reason for lack of knowledge is simply that it does not exist. Many problems are complex and few of them are well researched; and often the results of the research made are not available or are not operationalized.

10.3.3. Inadequate Information

Control-oriented medium- and long-term planning encourages the collection of large amounts of information *prior* to implementation, because they are needed to specify future activities in detail. In Tanzania the 1991 goal on which planning was based, and the rapid implementation that it would entail to reach it, have undoubtedly increased the urge to engage in huge information collection exercises in the pre-implementation period.[3]

This has created problems for several planning teams. One RWMP planner frankly admitted that his team had drowned itself in information. That particular RWMP was never really finalized (Ch. 6). The Danish planning team set out to collect data from each and every one of some 1500 villages prior to any implementation. This included information on population, livestock, settlement patterns, economic activities, crops grown, water quality and quantity, and so on.[4] On average, only two to three hours were available per village. The quality and validity of the data on individual villages may therefore often be problematic and seasonally biased.

A second problem with much of the information collected for planning purposes is that it quickly becomes outdated. None of the medium- and long-term plans included in this study have been updated on a regular basis. Only the Finns have now decided to fund the updating of the ten-year-old RWMPs for Mtwara-Lindi regions.

Information collected prior to implementation suffers from a third problem. It is often not operational. All of the plans studied lack, for example, implementation-oriented information on how the recipient bureaucracy works at various administrative levels. Mostly plans contain general laments about "inefficiency" and the need for "streamlining". There is rarely much information on and analysis of procedures, staff performance, carrier structures, organizational problems, conflicts, etc. Perhaps the political and bureaucratic sensibility of these issues make expatriate planners abstain. But lately such inquiries have been strongly advocated and to some extent practised (Cohen *et al.* 1985).

---

[3] There is a third incentive for planners to collect information – especially quantitative data.

> The expert's dependency on measurement is very real. Measurements and quantitative analysis are the bases of knowledge which differentiates them and, therefore, a basis of their social power. They cannot spend too much time talking in vague ways. Sooner or later they need to concentrate on the issues on which they can exercise their skills (quoted from Rondinelli, 1983, 6).

[4] Interestingly, the urge to collect information is mostly limited to *primary* data and information. It is rare, in the plans studied here, to find examples of extensive use of and reference to secondary data and information. Perhaps the reason is that the client is only prepared to pay consultants for the collection of information for which cost, extent and type can be defined in advance? Collection and analyses of primary data also require larger resources – an appealing prospect for consultants. A serious consequence of the bias for primary information collection and analysis is that little historical information is obtained. It is very rare, in the case study documents, to see analyses of past experiences with water development in the regions being planned for. Use of information from other regions and from previous plans is also very rare.

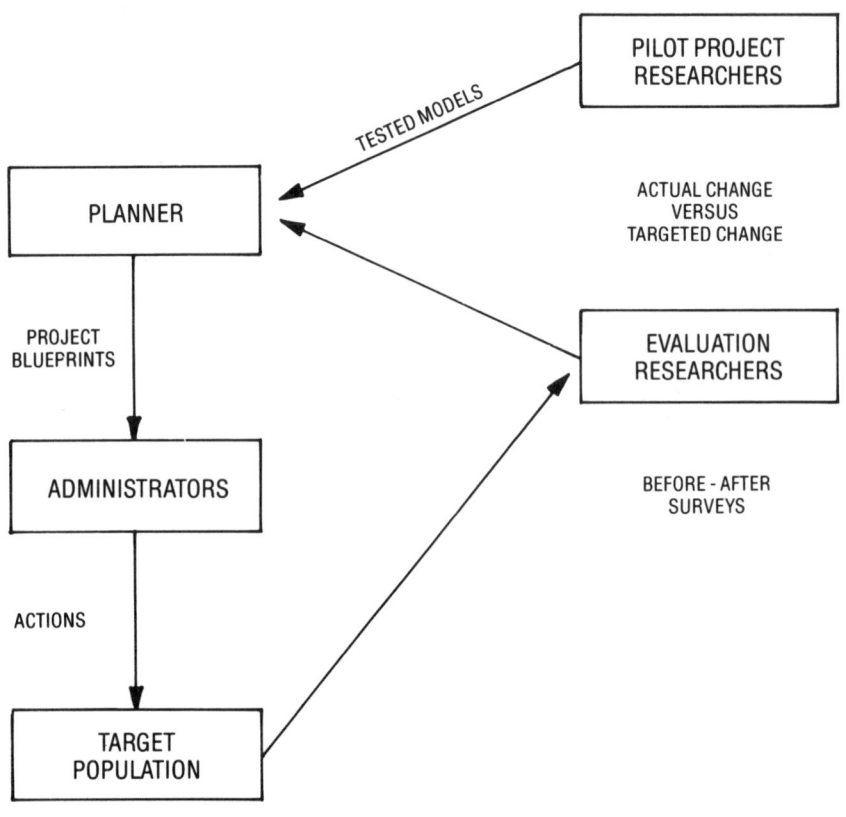

Source: *Korten (1980)*
Figure 4: *The separation of functions in control-oriented planning and implementation*

10.3.4. Planning without Implementation

A more fundamental reason exists why medium and long-term plans are typically made without the use of essential operational knowledge and information. It is that planning and implementation functions are organizationally separated. This is a key feature of the control-oriented approach (Figure 4). In the case of medium- and long-term planning for rural water sector development the sequence of activities has typically been as follows. Engineers, hydrologists, hydrogeologists and (sometimes) socio-economists carry out extensive information collection and baseline studies. On the basis of these, the planners (typically the same staff) prepare long-term plans for rural supply water development and/or specific short-term plans for implementation. Subsequently administrators and staff of the implementing organization are supposed to execute the project plan faithfully, much as a contractor would follow construction blueprints which indicate costs, time schedules, organizational arrangements, procedures, mechanisms for participation, etc. Well into the implementation period an evaluation team is supposed to measure actual versus planned changes and report this to the planners so that plans – subject to approval by the authorities – may be revised. Thus, the functions and institutional locations of the researcher-cum-data collectors, the planners, the administrators and the evaluators are sharply differentiated.

The result is that knowledge-building, decision-making and action taking roles are only integrated to a limited degree – if at all (Korten, 1980, 497). Researchers and planners do not have (and do not acquire) an intimate knowledge of implementation problems. They are separated from them. And administrators are stuck with plans for which they have had no responsibility and which may – in part or whole – be difficult or impossible to operationalize. Yet they can only change them with difficulty. Consequently, as Chambers (1973, 16) noted, "planning without implementation" often leads to "implementation without plans".

The separation of planning and implementation is characteristic of the cases studied in this research.[5] The only partial exception is the latter part of the socio-economic studies carried out in Iringa, Mbeya and Ruvuma (Ch. 8). Yet, the integration of planning and implementation has been identified as a very important factor in successful rural development projects and programmes (Sweet and Weisel, 1979; Korten, 1980; Moris, 1981; Paul, 1982; Morgan, 1983). This provides a strong argument for abolishing the strict separation of planning and implementation inherent in the control-oriented planning and implementation approach. More about this in Chapter 11.

10.4. Aid without Village-Level Roots

The need to involve users in rural water sector development activities has been strongly advocated both by Tanzanian authorities and by donors. Some pertinent statements are quoted in Chapter 2.4. But despite much rhetoric, the case studies show a remarkable lack of

---

[5] Indeed it is typical of most medium- and long-term planning exercises in East Africa (Chambers, 1973).

participation, particularly prior to 1983.[6] This is quite clear in the Finnish case where "participation" meant paying beneficiaries to do construction work. The Dutch case is an interesting example of the schizophrenia which is sometimes a convenient ailment for people involved in development assistance work. On the one hand, the consultants declared in a planning document:

> A crucial requirement of any shallow wells project is that it must be *people-oriented* rather than *technology-oriented* and involve the active participation of the village community right from the beginning (underlining in original; DHV (1980, 6)).

On the other hand, the same consultants were implementing a shallow wells project without any user involvement (see Ch. 5). The World Bank planners envisaged some degree of participation, but during implementation this was limited to a user payment for wells prior to construction (Ch. 7). In the Danish project, and now also the Swedish one, deliberate attempts at involving users in all phases of the project cycle are made. However, in both cases this participation is fairly strictly defined from above and it is too early to judge its outcomes in the long run (Ch. 6 and 8).

These general assessments will be exemplified below on the basis of the two dimensions of participation presented in Chapter 3.5: mobilization and empowerization.

10.4.1. Lack of Mobilization

Without user participation it cannot be assumed that the activities planned for a particular target group comply with their need. Yet in control-oriented planning and implementation the matching of activities and needs is typically taken as given. Although village priorities might be different, they are not considered during the planning stage and remain unknown to the planners until the implementation stage

---

[6] This general conclusion also holds for activities planned and implemented by the Tanzanian authorities without donor assistance. Although there are some reported cases of new water schemes being built on the basis of requests from village governments; of villages providing self-help labour during construction of schemes; and of villages repairing schemes on an ad hoc basis (Warner, 1973; BRALUP/CDR, 1982), there is no evidence of participation on a general scale of the mobilizing and empowerizing type envisaged by the Party since the early 1970s.

(see Ch. 10.2). A water scheme may therefore be built in a village that has other activities or improvements much higher on its priority list. It is a fairly typical problem, and some consequences have already been discussed (Ch. 10.1.3).

Control-oriented planning and implementation also tend to have negative effects on the willingness of users to commit resource to development activities. Yet such commitments are crucial from a participation perspective. They help to ensure that activities planned from above are indeed so high on the village priority list that users are willing to contribute some of their resources. User resource inputs (self-help labour, building materials, cash) also reduce the capital and/or recurrent cost to government and donors of rural water sector activities. Finally, contributions may increase users' sense of ownership of the projects. This is obviously crucial if users are to be responsible for sustaining a project in the future.

Unfortunately such resource commitments are less likely to occur when the control-oriented approach is used. Its emphasis on predictable production targets and fast-paced implementation schedules does not suit a participatory approach. The timing and magnitude of user resource commitments can neither be predicted nor controlled by planners and implementors. It often takes time before intended beneficiaries demand the services offered. Sometimes a demand does not exist at all. In such cases the control-oriented approach typically leads to a displacement of any participatory activities that may have been intended. The target group is simply provided with the planned service, as Chapter 7 on the World Bank experience shows. The other case studies also illustrate how short-term production consideration tends to replace participation. In fact, the whole notion of water to all by 1991 (or any other fixed date) is inconsistent with the notion of development based on beneficiary resource commitment. The pursuit of fixed targets leads to provision of services *to* people.

The third serious consequence of the control-oriented planning and implementation approach is that it is not based on beneficiary knowledge. With the strong emphasis on detailed pre-planning, and the separation of planners and implementors (see Ch. 10.3) the decision-making role is assigned to the individuals furthest removed from the intended users – the professional planners, the researches and regional- and central-level policy makers. Therefore, according to Korten, 1980 – 498):

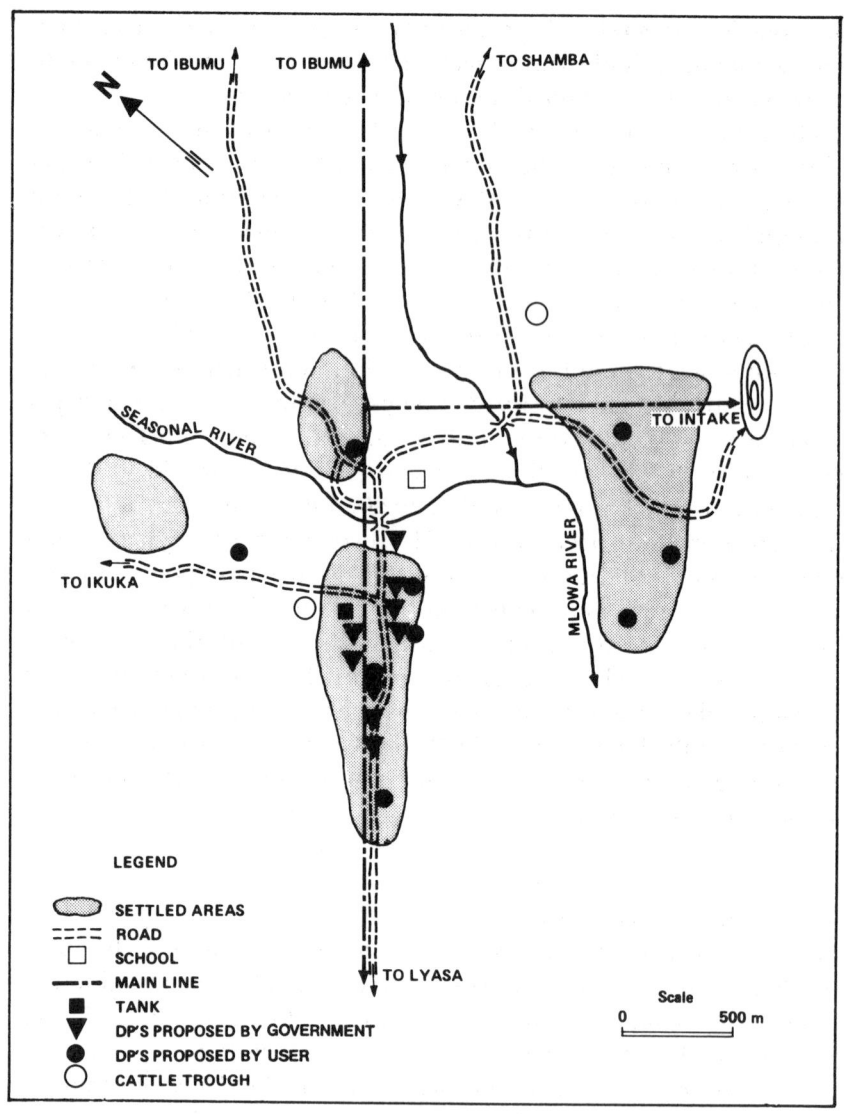

Source: *BRALUP/CDR (1982)*
Figure 5: *User versus Government Locations of Domestic Water Points*

> ... planners proceed as if they are writing on a clean slate and possessing all the knowledge relevant to improving the villagers' life. In reality they are making interventions into well-established socio-technical systems ...

Failure to solicit and use the knowledge of intended beneficiaries during both planning and implementation have several consequences.

Only one example related specifically to the *planning* of water schemes is presented here. It illustrates that beneficiaries can contribute significantly even to the technical design of water schemes.

The example concerns the location of water posts in one village in Iringa region. Figure 5 shows the locations of nine domestic water points made by the regional MAJI office, compared to the locations proposed by the village water committee. The locations of domestic points chosen by the committee generally match the settlement pattern much better than those proposed by MAJI. The locations chosen by the committee therefore result in a considerably higher service level. Although the total cost of the distribution system resulting from the committee's location of water points has increased, the cost per beneficiary has not. It decreased by approximately 20 per cent per person served.[7] The locations suggested by MAJI would lead to a smaller total cost. The reason is that no domestic points were located to the south-east of the River Mlowa. However, people living here would have continued to draw water from this occasionally polluted river. For very few drawers of water pass a traditional source to get to a domestic point. Much implementation has ignored this very basic point as some of the case studies exemplify. The example above therefore reflects a general problem – not the incompetence of particular MAJI staff. Chambers (1983), for example, has discussed the widespread professional bias against the use of local knowledge in development activities.

10.4.2. Lack of empowerization

The extent to which a process of "empowering" (Ch. 3.5) the rural population in Tanzania has taken or does take place has been much studied. The studies include analyses of participation through the Party (Cliffe, 1967); the Parliament (Tordorff, 1984); the regional and district administrations (Finucane, 1974; Mushi, 1978; Fortmann, 1980); the District Councils prior to 1972 (Dryden, 1968); and the cooperatives (Hyden, 1973). Without going into the details of these studies it is clear that the participation through these institutional structures is considerably less prominent than declared policies imply. It is also clear that donors operating in the rural water sector have limited direct influence on participation through these structures.

Moreover, in the projects studied, donors have made little explicit

---

[7] Several additional examples are given in BRALUP/CDR (1982, Ch. 6.4. and 6.7.).

effort to strengthen village level organizations and their capacity to influence project decision-making and eventually to take over key project activities (such as operation and maintenance). In the two cases where some efforts have been made (Ch. 6 and 8) it is too early to assess the outcome. It is, however, clear that the strengthening of local capacity and the enhancement of the role of women in local-level groups is a long and slow process for which the control-oriented planning and implementation process is not well suited.

10.4.3. Constraints on Participation

In discussing the limitations of control-oriented planning and implementation with respect to participation it is important to distinguish between those constraints which have political, structural, bureaucratic, economic or social roots and those which can be linked to specific planning and implementation approaches. Clearly a number of political constraints to participation exist both on the donor and the recipient side (Ch. 2.2 to 2.4 and Ch. 10.1). The byzantine planning and implementation structures described here are major impediments – as are the bureaucratic biases against rural people, that are widespread among both donor[8] and recipient staff. There may well be historical reasons why intended beneficiaries are simply not very interested in participating in government-initiated and donor-supported water activities (the forced movement of people into village settlements; the strong pressure on peasants to "participate" in communal farming etc.). Finally, participation may simply fail because not enough money and resources are allocated to that staff which is supposed to help implement it. The combined effect of these factors is not conducive to participatory planning and implementation in Tanzania (Holmquist, 1979). But this context is rather typical for rural development in many developing countries.[9] The donors' use of the control-oriented planning and implementation approach tends to accentuate the effects of the contextual constraints, as discussed above. Participation is therefore a concept in need of an implementation strategy. In Chapter 11, two alternative but participation-oriented planning and implementation approaches are presented.

---

[8] The donor staff bias against participation is common according to Korten (1980, 498) and Chambers (1983, Ch. 7).
[9] Gow and Van Sant (1985) have reviewed some of the relevant literature.

10.5. Aid without Institutional Roots
Bypassing of local institutions is a logical consequence of the control-oriented planning and implementation approach. When a donor circumvents parts of the recipient institutions and procedures, the aim is to gain better control over planning and implementation. The typical tools for this have been independent project units (Ch. 4–5); fairly autonomous sections within existing organizations (Ch. 7–8); and coordination bodies separated from existing regional and district coordinating committees (Ch. 4.10 to 8.10). Such bypassing is widespread in much development assistance (Honadle et al. 1983b).

The degree of bypassing is difficult to measure. It varies from donor to donor and over time as the case studies show. Some degree of bypassing is, however, characteristic for all the cases included here. During the preparation of medium- and long-term plans the bypass has been almost complete to the extent that many expatriate planning teams had little contact with local institutions and included only a few counterparts or none at all. During implementation this separation continued in certain regions, while other donors preferred a larger degree of integration of their activities (Ch. 4.3. – 8.3).

One clear result of bypassing has been the fast production of new schemes and wells in several donor projects (notably in the Finnish, the Dutch and to some degree the Danish). No doubt production – and the propensity of donors to fund it – would have been significantly smaller had donors been without the control. These short-run gains have considerable costs, however.

Bypassing results in a strong bias towards projects rather than programmes in development assistance. The project bias in turn results in fragmentation. Projects are well suited to execution by autonomous or semi-autonomous units (Korten, 1980, 484). Like such units, projects are terminal. They do not commit the donor indefinitely. Projects also limit donor commitment to the funding of preplanned specific activities, facilities and equipment. As already mentioned, the number of projects in the rural water sector is substantial (Ch. 10.2). This project proliferation puts a significant burden on recipient organizations. A continuous stream of donor teams is sent out to negotiate, plan, implement, monitor and evaluate activities. Each donor works according to its own arrangements; requests senior government officials to join in this work; and demands high-level Ministry staff to function as liaison for *their* projects. At present, for example, three to four senior Ministry officials serve as "link-men" to donors' rural water projects. The officials are placed in different divisions in the

Ministry and do not appear to coordinate this work among themselves. Donors tend to compete for whom they consider "best", and to avoid the younger staff that are officially appointed for this function. A similar lack of local control and overloading of local administrations can be observed at the regional level. Here the number of donors is smaller but the regional administrations are also significantly weaker. Finally, it should be noted that fairly autonomous project units tend to confuse and complicate accountability for plans, construction of new schemes, and their operation and maintenance. For instance, all consultancy firms preparing RWMPs had contracts with the donor – not the recipient (Ch. 2.2). The same is the case – with one exception – for those consultants involved in implementation of rural water projects. Thus, the Ministry of Water does not have direct authority over the consultants. It is spread among several donors. The end result tends to be *fragmentation* of recipient institutions. Another clear example of fragmentation related to accounting is given below.

Bypassing, furthermore, tends to create dependency and hostility. Donor projects are often perceived by adjacent or redundant administrative and political units as a threat (Moris,1981). An excessive bypassing of line ministries and local authorities also creates jealousies and distorts priorities that might later undermine project implementation. Recent reports by the Ministry of Water and Energy reflect emerging hostility towards donors at the central level. In its review of the Morogoro shallow wells project, for example, the Ministry of Water and Energy (1984, 4.11) concluded that "we have to decide what we plan to do rather than being told by donors and contractors." The Dutch shallow wells project provides another example of hostility at the regional level (Ch. 5). And regional politicians in Iringa, when interviewed, expressed concern over what they regarded as their limited influence on donor-funded activities in their region: "Donors try to avoid us." Jealousies are undoubtedly exacerbated by the special resources controlled by donor-funded projects and by the special claims on local personnel that donors often make. For this helps to create projects which are oases of plenty in resource-starved surroundings. In turn this creates new opportunities for corruption.

The view that nothing can be done within the system is also amplified when bypassing occurs. It perpetuates dependency syndromes (Honadle and Rosengard, 1983). Strachan (1978) argues that bypassing also leads to passivity in counterpart institutions. When major decisions and plans are continuously made by the donors and their consultants, he claims, it contributes to a "lack of initiative" and to

Table 12. *Direct and Total Costs of Wells Construction in Donor Assisted Projects*

Implement through	Cost to project per person served (Tsh)			Cost figures from
	Direct[a]	Total[b]	Factor	
Finnwater, Mtwara-Lindi	60	160	2.7	1978–83
DHV, Morogoro	70	250	3.6	1978–81
World Bank, Mwanza	120	360	3.0	1979–83

a Labour, transport, materials and direct overhead
b Direct cost plus cost of consultants, housing, vehicles, etc.
Sources: Mtwara/Lindi: FINNIDA (1984, p. 13 and Annex 8, p. 2)
　　　　 Morogoro: Hordijk *et al.* (1982, ch. 6.5)
　　　　 Mwanza: PEU (1984a, Tables 2, 6, 7 and 9)

"subterfuge". Some of the cases studied here (notably Ch. 4 and 5) support this observation.

The cost efficiency of bypassing – even the short term efficiency – is easily overrated. For increased outputs should be related to the actual cost of donor-controlled implementation. This is difficult to do because of the fragmentation of accounts. They are rarely broken down into the relevant categories. Nor are all relevant costs recorded in the same accounts.[10] Cost efficiency comparisons are also difficult because of differences in standards, technical quality, water resource potential, etc., in the various regions. Nevertheless, available figures on wells construction costs shown in Table 12 suggest that the cost of consul-

---

[10] Typically, they are spread on three accounts: a MAJI account for salaries, etc. of staff seconded to the donor project; a donor project unit account; and a donor headquarters account for costs of directly employed donor experts, overseas purchases, etc. Not only does this fragmentation make donor cost-control difficult; it also plays havoc with recipient accounts. Thus the Auditor-General (URT, 1981, 59) wrote:

> ... in the absence of proper supporting documents with full details of the aid given, it could not be established in audit that all direct-to-project aid actually received during the year had been adjusted in the accounts.

Sometimes the lack of coordination between recipient and donor accounts contributes significantly to the huge differences between actual and budgeted capital expenditures (see URT, 1981, 195).

tants, experts and housing, etc., are considerable. This leads to very high total costs of water supply improvements. The table indicates that the total cost of wells construction is around 300 per cent larger than the direct costs. Much of this differential is accounted for by the cost of the institutional bypassing arrangement and the dominant role of expatriates that it entails. For comparison it should be noted that the total cost of MAJI-constructed schemes is thought to be between 60 per cent (WHO/World Bank, 1977, 7) and 100 per cent (Hordijk *et al.*, 1982, 70–71), but these figures are very uncertain.

It could also be argued that some of the overheads include the cost of institution-building efforts, such as training and the development of new procedures to strengthen counterpart institutions. However, in the cases studied, the technical assistance staff has only spent limited time on such activities.

This leads to the final problem with bypassing. It diverts attention from the institutional development activities mentioned above. In fact it diverts attention from the post-project period when the donor has withdrawn. Instead, attention is focused on the project period itself. There are several indications of this in the case studies: the limited number of counterparts involved in both planning and implementation; the low priority given to training of MAJI staff at all levels; the preference for establishing specific project rules rather than attempting to make necessary changes in recipient procedures; and the lack of a long-term strategy and provisions for transferring project activities to recipient institutions. The neglect of these factors has clearly contributed to the non-sustainability of the Dutch and the World Bank projects. More than 10 years after its inception, the Finnish project is realizing that its presence in Mtwara-Lindi regions is needed for an unspecified future period because of the important factors neglected in the past. The Swedish and Danish projects are still fairly new, but the Danish one has become increasingly production-oriented during the past couple of years. Non-sustainability is therefore a general problem in the donor-assisted projects. And, as argued throughout this chapter, it is not caused by host country problems alone.

There are several reasons for bypassing. The control aspect itself is one reason, as has already been mentioned (Ch. 10.1). The limited capacity of recipient organizations to plan and implement is another reason. And this capacity has decreased steadily over the last ten years as a result of the economic crisis (see Ch. 2.3). Instead of adjusting planning requirements and implementation targets to existing local capacity, donors and recipient chose to substitute the perceived deficiencies in local organizations with technical assistance organized mainly outside existing organizations. A third reason for bypassing is clearly related to the notion of efficiency. It is assumed that technical assistance staff – if allowed to work more independently of local organizations and bureaucratic red tape – can work so efficiently that this justifies their cost. The efficiency motive is obviously important, considering the implementation pressure that all donors projects but one (Ch. 6) have been under. Finally, the increasing inefficiency and corruption in the Tanzanian public sector has been a strong, but unmentioned, motive for bypassing. It has undoubtedly been felt that donor assistance had partly to circumvent local organization if it was to produce results and reach the villages as intended.

Underlying all these motives has, finally, been the implicit assumption that some time in the future the various activities undertaken through the donor-controlled organizational structures could be transferred back to recipient institutions so that the benefits of these activities could be sustained.

On balance, however, the short-term benefits of bypassing do not appear to cancel out its various negative aspects in the cases studied. Although some degree of bypassing may be justified under certain conditions it has generally been too pronounced in the donor-assisted water projects studied here. In Chapter 11 alternatives to bypassing are presented.

10.6. The Limits: A General Trend

Various assumptions of the control-oriented approach have been analysed. Their validity is often questionable. This contributes to a number of problems. Among them are various degrees of non-use of medium- and long-term plans and non-sustainability of donor-supported activities.

The cases studied in this report concern rural water supply development in Tanzania. The frequent references to the literature indicate, however, that the problems of the control-oriented approach are general and widespread. They occur at different levels, in different sectors, and in different developing countries (see Chapter 9.3). They are so well documented that they amount to a crisis in development planning. Two well known summary studies are very explicit on this. Thus in their now classical study of planning and budgeting, Caiden and Wildawsky (1974, 293) came to the following conclusion:

If we were asked to design a mechanism for decision to maximize every known disability and minimize any possible advantage of poor countries, we would hardly do better than comprehensive multi-sectoral planning .... For this calls for unavailable information, non-existent knowledge and a political stability in consistent pursuit of aims undreamed of in their existence.

Korten (1980, 495), in his well-known study of five successful rural development programmes, is equally certain:

> ... if success of any such (programme) was an outcome of project papers, social benefit-cost analysis, environmental impact analysis or PERT charts, the source documents examined made no mention of it.

It is for these reasons that alternatives to the control-oriented planning and implementation approaches are needed. They are presented in the final chapters.

Part Four
Alternatives

11. Rural Water Supply Development through Experiments and Learning

The general conclusion then of this study is that control-oriented plans have not been very valuable for rural water development. Chapter 9 provides a summary of the evidence and Chapter 10 discusses some important reasons. This study therefore reflects the doubt and confusion which increasingly exist with respect to development assistance as practised by donors, and development administration as practised by recipient governments (Blair, 1985, 449). It follows a long period of confidence that development assistance could be planned and managed according to western models and that such models could and should be transferred to and used by host country bureaucracies. But the gradual shift from heavy infrastructure-type activities towards basic needs activities – together with the deepening crisis in many developing countries – has helped to scatter this confidence.

In their extensive review of rural development planning and implementation, Johnston and Clark (1982, 219) are quite adamant in their criticism of control-oriented planning and implementation:

> This approach has been totally and repeatedly discredited by experience; yet despite much rhetoric to the contrary, it persists at the conceptual core of much development activity.

How may the situation be improved? It should be obvious from the arguments elaborated in Chapter 10 that the answer is *not* necessarily to insist on more resources and time for more comprehensive planning and conventional research of development activities prior to implementation. It is the underlying assumptions of the control-oriented planning and implementation which are at fault. Unfortunately, the Tanzanian rural water sector has no clearcut "success-stories" on which proposals for improvements of planning and implementation could be based. It is necessary to turn elsewhere.

New approaches are now being offered under various labels, of which the most commonly used is "bureaucratic reorientation" (Blair, 1985). Changes in planning approaches are just one element of this reorientation. The basic idea is a simple one – yet antithetical to most bureaucratic tradition. It is to make donor and host country bureaucracies responsive to the intended beneficiaries of development activities and to their own lower-echelon staff. Neither trait is prevalent today. To bring about such reorientation will not be easy at all (see Chapter 10 and Chapter 11.3). For bureaucratic reorientation necessitates, among other things: (i) adopting a participatory learning-based approach to planning and implementation in which things are expected to go wrong but where mistakes are seen as opportunities to improve projects and programmes; (ii) decentralizing planning and implementation and providing incentives for donor and host bureaucracies to engage in the learning-based approach; and (iii) proceeding gradually so that the knowledge required to plan development activities is gained simultaneously with the capacity to implement plans. Thus as projects or programmes mature, they may grow larger. Eventually more specific and detailed planning may be both possible and necessary.

Before proceeding, however, it is important to clarify some potential misunderstandings. The proposed approaches are not relevant for large capital-intensive infrastructure projects nor for activities that have run successfully for a number of years. They are geared to rural development activities in which the participation and resource commitments by target groups are crucial for success, such as activities related to water supply, health, peasant agriculture, agro-forestry, etc. They are also relevant when there are major uncertainties about such issues as, for example, objectives, user needs and reactions, appropriate technologies and methods, institutional capacities, political support or resistance. Finally, the proposed approaches are not of the recipe type. Their specific application would vary substantially depending on purpose; context; whether government or private; whether a new or established organization; whether locally- or donor-funded; and the particular learning stage reached by the programme or project (Korten, 1980, 502). *However, the essential nature of the process would remain much the same.*

Two examples of the most relevant alternative approaches to planning and implementation of rural development activities are given below. There are substantial similarities between them.

11.1. Development Projects as Policy Experiments

Rondinelli's (1983, Ch. 4) starting point is simply to recognize some key characteristics of rural development activities and the context in which they occur: the complexity and uncertainty, and the corresponding lack of specific knowledge on which to base extensive pre-implementation planning. Consequently, he regards all development projects as policy experiments. Planning should therefore be incremental and adaptive. Problems should be disaggregated and responses formulated through a process of decision-making that links learning with action. And this process should be based on past experiences.

The proposed framework is depicted in Table 13. It consists of a four-stage process of project planning and implementation that seeks to cope with problems in an experimental, incremental and adaptive fashion. The figure directs attention to issues about which little is known or about which there is usually a great deal of uncertainty. It also shows some of the main considerations about how to proceed, as well as the major risks involved.

Experimental projects are generally small-scale, highly exploratory and risky. Frequently they do not provide direct or immediate results. They require sheltered and privileged conditions (autonomous organization, generous budgets, and substantial input of highly skilled staff and researchers). They are relatively expensive. They are justified when there are major uncertainties. They cannot be planned in a conventional way. Their main benefit is the knowledge which is generated – both if they fail and if they succeed.

Pilot projects are basically used to test the experiences gained in experiments; to test their applicability in places with conditions similar to those under which experiments were performed; and to test the feasibility and acceptability of the experimental findings in new environments. Pilot projects should be more exposed to real life conditions than experimental projects. Their funding should be less generous. Use of foreign personnel and imported equipment should be limited. The project may not have to be autonomous but may be partly integrated into the existing organizational structure. Users should be actively involved. Planning should be kept flexible and responsive, for there are still major uncertainties. The major benefit of pilot projects is that they help to indicate the acceptability and feasibility of existing knowledge in new locations or contexts.

Demonstration projects are initiated when pilot projects have turned up

Table 13. *The Experimental Approach*

Stage	Experimental	Pilot	Demonstration	Replication
Why	To obtain knowledge about unknowns and problems	To test acceptability and feasibility of existing knowledge in specific contexts	To demonstrate that new technologies, methods, and programmes are better than presently used ones	To expand productivity and administrative capacity to disseminate and deliver
When	Major uncertainties about: – objectives – alternative solutions – methods of analysis or implementation – appropriate technology – required inputs – adaptability to local conditions – transferability or replicability – acceptability by users – dissemination or delivery systems	Major uncertainties about: – methods of analysis or implementation – appropriate technology – adaptability – transferability – acceptability – dissemination or delivery systems	Major uncertainties about: – replicability – acceptability – dissemination or delivery systems	Major uncertainties only about: – dissemination or delivery systems on large scale
How	– Obtain high-level political support – high input of specialized professional staff	– same; plus local political support – mostly existing staff, committed leaders	– same – same; extensive training of field staff in techniques and participatory methods	– same – same

192

– autonomous organization	– semi-autonomous organization	– less by-passing of existing organizations; their capacity to be increased gradually	– less by-passing of existing organizations, coordination with other organizations
– sheltered, flexible and generous budget	– sheltered, flexible budget limited to resources available in ordinary programmes	– partly sheltered	– ordinary budget item
– close dialogue with users	– active participation by users; two-way communication between staff and users	– immediate and direct benefits for users; build-up user capacity to participate and control	– same
– activities limited to few locations	– activities gradually expanded in scope and space	– locations most favourable for success selected first	– expansion requires local adaptions
– general guidelines for activities and strategies	– flexible and responsive planning	– standardized procedures; make more specific plans establishing regular monitoring	– same; emphasis on programming and scheduling
Major risks – necessary political support withdrawn or excessive	– same; plus overt opposition of political leaders, interest groups or bureaucrats	– same	– same
– frequent failures; cost paid by sponsors	– same	– cost of failures shared by sponsors and users	– same
	– too much tried too fast	– overemphasis on expansion prevents changes in activities necessitated by going to scale	– adaptive approach harder to implement when activities run on large scale

Source: Summary of Rondinelli (1983, Ch. 4)

useful experience on the basis of which new technologies, methods or programmes can be made. Their aim is to show that these new approaches are better than traditional ones because, for instance, they lower production costs or deliver social services more efficiently. The implementation of demonstration projects may be integrated further into the existing institutional framework than during the previous stages. Major emphasis in this stage is on capacity-building through training. Field-level staff and beneficiaries are especially important target groups for this. More specific planning may now be made and procedures may be standardized based on previous experience. Regular monitoring should be established. However, major adjustments may still be needed – even at this stage.

Replication projects are the final stage in the experimental series. The main purpose of these projects is to expand productive and administrative capacity to implement large-scale production or service delivery. By now the project might be fully integrated into existing structures and it should become an ordinary budget item. Organizational problems become paramount, because even when technical, administrative and socio-political uncertainties have been resolved in earlier stages, new problems tend to arise simply with the expansion of scale. Bureaucratic responsiveness to user needs (these may vary significantly in areas covered by large-scale projects) remains crucial. Planning and programming can now be done in detail, but adaptations might still be necessary.

Throughout the stages there are major risks of failure. Some of those risks may be expected to diminish as projects progress through the four stages – failures due to inappropriate technologies and implementation methods; resistance or lack of interest from users, etc. After all, reduction of such risks is the main rationale for proceeding gradually. Yet risk of failure for political reasons remains high throughout the four stages. They may originate from within the project (staff resist the changes being introduced) or from without. The latter can take many forms. Political support from donor and host for certain projects might be so strong and excessive that rapid expansion through the stages is forced before enough is known to reduce uncertainties or before the organizational capacity to expand exists. Political resistance might obviously also occur – even if projects prove to be technically, socially and economically feasible. Pyle (1980), Sussmann (1980) and Quick (1980) have provided detailed evidence about the politics of going from experimental projects to larger-scale expansion.

Regarding development projects as policy experiments obviously does not square well with normal donor and host country approaches. They prefer pre-implementation planning that specifies inputs and outputs and timing of activities in detail. Implementors then have to stick to these plans. The new approach requires planners and administrators

> ...to view social problem solving as an incremental process of social interaction, trial and error, successive approximation and social learning. Such an approach requires an institutional context quite different from that of Weberian bureaucracies. (Rondinelli, 1983, 128)

Adopting this experimental approach to planning and administration requires a "reorientation of development administration", according to Rondinelli (1983, Ch. 5). First, authority for planning and implementation must be decentralized so that it may become more responsive to beneficiary needs. Strong central political support is needed to initiate this, but at the same time it cannot be implemented and sustained without widespread participation and political support from below. Second, therefore, is a need to "create" countervailing power among rural people to force bureaucracies to respond to their needs. Sometimes this requires breaking the hold of clientist policies and power. Often it necessitates the "creation" and "empowerization" of an organizational base of political support and local participation. Third, planning and management procedures must be made simpler. Complex methods rarely work at any level of government in developing countries, especially not at the local level in rural areas. Rural people either tend to ignore such methods or are manipulated by officials who use them. Fourth, comprehensive and control-oriented planning should (also therefore) be replaced with adjunctive and strategic planning. This is aimed at (i) facilitating decision-making among a wide variety of organizations and interests in society; (ii) analysing courses of action that produce incremental changes in existing conditions; (iii) converting large, complex development problems into smaller disaggregated ones that can be dealt with incrementally; and (iv) focusing on examining goals which can be reached with resources likely to become available and which reflect the interests of those groups which participate in or benefit from projects. Fifth, the attitudes of staff in host and recipient bureaucracies must change. Engaging in a learning-based approach to planning and implementation obviously moves much of the action from the splendid isolation of the desk to the turbulent conditions in the field. The changes in organizational and personal attitudes implied in this may be difficult enough to achieve. Undoubtedly it will be more difficult to change administrators' attitudes to errors. Failures, mistakes, continuous adaptation and redesign are inevitable in the experimental approach. Yet donor and host bureaucracies often suppress errors and sometimes punish staff when mistakes are made. This inhibits creativity, flexibility and innovation – the very stuff of successful project planning and implementation. Finally, according to Rondinelli, administrative systems and career possibilities should be changed to reward these attributes without which host and donor bureaucracies remain rigid, unresponsive and control-oriented.

Rondinelli's proposal is valuable because it provides a systematic approach to planning and implementation under conditions of uncertainty and complexity. However, these conditions are likely to keep changing. Even in the replication stage considerable adjustments may therefore be needed. Perhaps Rondinelli's concept of the replication stage is too static.

11.2. The Learning Process Approach

Korten's (1980) starting point is essentially the same as Rondinelli's. Their proposed approaches are also similar in many respects. However, where Rondinelli's approach is project-oriented Korten's is programme-oriented. Korten argues that the project concept itself and its emphasis on breaking development into discrete, time-bounded pieces might be the real heart of the problem:

> In rural development few important outcomes are terminal . . . Constructing an irrigation system is terminal. Improving and sustaining efficient, reliable and equitable access to water is not. [The numerous] underutilized irrigation systems which serve only a fraction of their designed service area . . . serve as sober testimony to the limitations of a terminal approach to development. (Korten, 1980, 508).

Thus Korten insists that beyond initial experiments, new development activities must be developed in an adaptive, bottom-up process of programme and organization development through which an adequate fit may be achieved between beneficiary needs, programme outputs and organizational capacity (see Figure 6). This calls not for more sophisticated skills in the preparation of detailed plans, but rather for skills in "building capacity for action through action".

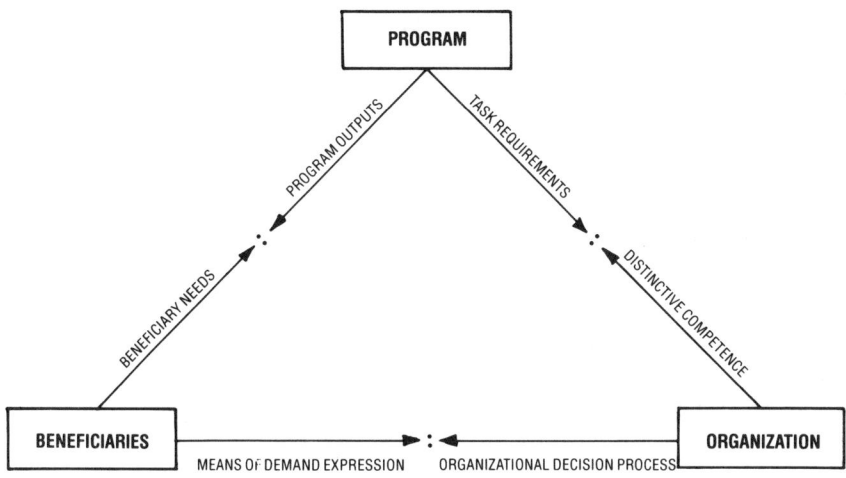

Source: *Korten (1980)*
Figure 6: *Schematic Representation of Fit Requirements in the Learning Approach*

Korten suggests that this may take place in a learning process that proceeds in three stages.

In stage 1 – learning to be effective – efforts should concentrate on developing a programme model that matches beneficiary needs. Secondly, it is critical to find a fit between how beneficiaries can define, communicate and insist on their needs and the process by which the organization makes decisions. This may require support for village level organizations that can express village needs and demands. New procedures which make it possible for the organization to respond to such demands are often required too. The way this is – or can be – done will largely determine to what extent the programme will strengthen or diminish community control over programme activities. It is an experimental situation requiring a relatively large input of highly skilled staff and substantial freedom from normal administrative constraints. Researchers might be needed, but they should work together with staff from the programme organization and with local groups in learning what is required for a given time and setting. As in most learning processes, this stage might involve basic learning about community dynamics and conflicts and simply the asking of relevant questions. This action research is relatively costly; many errors will be made, and efficiency will be low. The products of the action team are not necessarily a conventional research report but rather proposals for new methods, procedures and training approaches. Transition to stage 2 begins when the programme output is found to be effective in responding to an identified village level need and when this output matches the capability of the action research team to supply it. Korten claims that only one or two of ten programmes funded for stage 1 may proceed to stage 2, and that stage 1 typically lasts three years or more.

In stage 2 – learning to be efficient – the general focus is on reducing costs and eliminating activities that are not essential for the maintenance of effectiveness. This entails (i) achieving a match between programme requirements and organizational capacities under the constraints which must be expected if the programme expands (budgets, for example, should typically be lean as too much money too early will kill the learning process); (ii) trying out and refining the working model in additional locations through modest expansion; and (iii) gradually increasing the core staff involved with the programme, standardizing procedures and developing supporting management systems. Once these tasks have been achieved with acceptable levels of effectiveness and efficiency, transition to stage 3 can begin.

In stage 3 – learning to expand – the focus is on an expansion of the programme. A more detailed plan and budget for this expansion should be prepared, based on the experiences and capacities developed during stages 1 or 2. The emphasis in stage 3 will be on expansion of organizational capacity, but continued refinements of programme output to meet beneficiary needs and village demands and commitment may still be needed. This requires that flexibility be maintained, with close attention to what happens on the ground. The amounts of money moved is still limited. The rate of expansion will be determined by how fast the necessary organizational capacity can be developed. By the end of stage 3 the programme should have matured to the point of a relatively stable large-scale operation. It may take ten years or more to get that far. Only then is a programme able to absorb substantial amounts of money. And only then may it be fit for detailed programming.

Obviously the structure of each stage will depend on the specific setting of the programme. Also important is whether a new organization is being built up from scratch or whether the task is to build an appropriate new capacity in an existing organization. However, the basic features of the processes involved will be much the same. Programmes and the organizations supporting them "evolve and grow". They are not "designed and implemented".

Korten also points to the need for bureaucratic reorientation. He advocates "learning organizations" with a capacity for responsive and anticipatory adaptation which must embrace error; planning with the intended beneficiaries; and linking knowledge-building with action.

Learning organizations look on errors as a vital source of information for making adjustments to achieve the needed fits described above. They do not deny errors, nor are they paralyzed by them. Instead, they are characterized by the candour with which errors are discussed and analysed and by the corrective actions they are undertaking. Such organizations develop procedures for monitoring field activities that are simple and which encourage field staff to report what they actually see and do, and not what they think the management would like to hear.

Learning organizations plan in close cooperation with the intended beneficiaries. This is based on the premise that rural people have a great deal to contribute to programmes. Their knowledge can be crucial to any effort by governments to carry out development activities. Building on, rather than replacing, indigenous knowledge reduces the risk that programme intervention will "de-skill" the villagers and thus increase their dependence on external experts and bureaucracies over whom they have little, sometimes no, control.

Finally, learning organizations gain knowledge through action. Those persons in close contact with village reality – the administrators, the field staff and the villagers themselves – cooperate with researchers and planners to define village needs and programme output. Especially in the early stages, the roles of researchers, planners and administrators are combined within the same team. Even as programmes grow this integration is stressed. Researchers work hand-in-hand with operating personnel, planning is done by those responsible for implementation; and managers spend a substantial amount of time in the field keeping in contact with village reality.

11.3. The Adaptive Approach

Rondinelli and Korten present many of the salient features of the proponents of new approaches to rural development planning and implementation. Others have elaborated on the same theme of bureaucratic reorientation, adaptive planning and implementation (Moris, 1981; Paul, 1982; Johnston and Clark, 1982; Chambers, 1983). It is easy to criticize them since, on the one hand, they all emphasize the political nature of planning and implementation and point out that conventional approaches basically ignore this. On the other hand, the alternative adaptive approaches proposed by them need a considerable "political will" in order to become operational. Yet lack of this particular will is mainly a reflection of political conflicts. An instant change in donor and recipient approaches to planning and implementation from a control-oriented to an adaptive one should therefore not be expected. Less will do.

12. The Adaptive Approach to Planning and Implementation of Rural Water Supply Development in Tanzania

Rural water sector development takes place in a context which is characterized by conflicting aims between multiple decision-makers; by a lack of predictability, knowledge and information about many crucial sector activities and their environment; by a need to utilize user knowledge and local resources and to empower beneficiaries to participate; and by a need to strengthen the capacity of recipient organizations to carry out activities on a sustained basis.

As argued in Chapter 10, the control-oriented planning and implementation approaches used so far by donors and recipients in the rural water sector build on assumptions and lead to practices which are not appropriate in this context. The adaptive approach as presented in Chapter 11 appears more promising given the key characteristics of the task environment analysed in this study.

It is therefore worthwhile considering alternative ways of structuring donor involvement in the Tanzanian rural water sector compared to the approaches used during the 1970s. The proposals presented below are based on (i) the obvious benefits of hindsight; (ii) certain optimistic assumptions about the willingness of donors and host to engage in some bureaucratic reorientation; and (iii) selective use of the proposals by Rondinelli and Korten to fit the *ongoing* rural water projects and programmes in Tanzania. Three main issues need to be addressed: sector policies; planning and implementation methods; and institutional capacity.[1]

---

[1] Rondinelli (1983, 106–148) has provided a detailed description of eight factors to consider in the introduction of an adaptive planning and implementation approach.

12.1. Sector Policies and Donor Assistance

It should be possible to learn from the experience of the last 15 years and from the errors committed by both recipient and donors. Unfortunately, retrospective analyses are conspicuously absent from all the medium- and long-term plans included in the case studies. They have, instead, been squarely fixed on the future. Likewise, the water sector policies of 1971 (Ch. 2.4) have basically remained unchanged ever since. A sector review is therefore clearly needed. The last one was made in 1977.[2] The aim of the review should be to formulate a set of long-term policies which can provide a new framework for planning and implementation at various administrative levels. These policies should focus on processes rather than targets, on principles rather than specifics, and should be based on an assessment of present sector policies in their local, institutional and macro-economic context. From the village level and upwards the key issues that need to be considered to formulate new policies are:[3]

– the degree of village ownership of completed water schemes;
– the role of beneficiaries in cost-recovery;
– the roles of the new district councils in rural water supply development in relation to village and regional/ national authorities (especially with respect to O & M activities).;
– the roles of the Ministry in sector activities (especially with respect to policy-making and coordination of donors, procurements, etc.);
– the roles of other recipient organizations in sector activities;
– the rough magnitude of local and donor funds likely to be available for capital and recurrent expenditures in the future;
– the priority choices between construction, rehabilitation and operation and maintenance activities;
– the identification of necessary initial changes in procedures, organizational structure, and staff to introduce a more adaptive planning and implementation approach;
– the priorities with respect to strengthening recipient organizations to carry out the new policies;

---

[2] Only two sector reviews have been made so far, by Rimer (1970) and by WHO/ World Bank (1977).
[3] The list does not include technical issues such as choice of technology, materials and equipment; water resource data; etc. They have already been extensively analysed and discussed over the last 15 years. The RWMPs provide much valuable information on these issues. In fact the best future use of the RWMPs appears to be to utilize their water resource data to establish a nationwide resource-monitoring system.

– the identification of the need for donor finance and technical assistance to implement the above, including a framework for better donor-recipient coordination.

It would be a key task in this review to arrive at a priority list of the changes to be made. Across-the-board changes are unlikely to be acceptable or implementable. A new set of national policies is needed. It is such policies which should provide the framework for adaptive planning and implementation. It could be said that the need for clear sector policies in adaptive planning and implementation corresponds to the need for detailed medium- and long-term plans in the control-oriented approach.

In principle the Ministry itself should make such reviews on a regular basis. Only recently has it begun to do so in a critical and analytical way (Msimbira, 1984; Ministry of Water and Energy, 1984). Looking back, it is clear that some of the resources spent by donors on medium- and long-term planning could have been better spent on strengthening the capacity of the Ministry to do recurrent sector and policy analyses. Given the likely heavy dependency on donor assistance to rural development in the future, a combined recipient-donor review appears appropriate at present.

There is no reason to be too "optimistic" about the impact of such reviews, however. Differing agendas will still persist among the many decision-makers (Ch. 10.1–2). But at present a considerable confusion exists both among recipient and donor institutions. The sector policies of 1971 ("free" water to all by 1991) are clearly unrealistic but they have not been replaced. Perhaps it takes a number of major crises, such as those discussed in the case studies, to force adjustments in the thinking of both donors and host. Whatever new strategies emerge from these adjustments should be regarded as temporary emphases in the learning process rather than unchangeable decrees.

12.2. Introducing Adaptive Planning and Implementation

Hopefully a joint recipient-donor sector review may clarify at least some of the confusion which exists at present about crucial economic, political and institutional issues. Such clarifications will obviously increase the benefits of the introduction of adaptive planning and implementation approaches. But even without them, parts of the new approach may still be worth introducing. For at present, as illustrated in the case studies, the implementation in the regions takes place without much use of the RWMPs and without a common strategy.

To be effective the adaptive approach requires (i) an information system that provides relevant, timely and succinct information; (ii) an organizational structure in which the role of the planner, the implementor and the evaluator are no longer sharply differentiated; and (iii) a set of procedures that allow a continuous dialogue with and involvement of the target groups to ensure that local needs are met and beneficiary knowledge and resources are solicited and used.

An *information system* for continuous monitoring and evaluation is very important. It should be related to the programme objectives in each particular region and such objectives should be much more explicitly and operationally stated than they have been hitherto.[4] Few people would disagree with this. Even proponents of control-oriented planning and implementation would not. Yet the case studies show that no such system has been established in any of the donor-assisted projects studied. Neither Tanzanian nor donor authorities know, for instance, the extent to which individual schemes *function*. Such crucial information is only collected on an ad hoc basis – if at all. Collection of information on the *utilization* of schemes by users is even less common. It was obtained for planning purposes by a few RWMP teams (see, for instance, Ch. 8), but appears to have been totally ignored during implementation.

Even information which would appear to be of key importance to control-oriented planners and implementators is not systematically collected. The lack of cost data on the construction and maintenance and operation of individual schemes is a good example. The lack of data on the real cost of technical assistance on donor-assisted water projects is another. The fragmented accounting systems add to the problem (Ch. 10.5).

It means that both the donors and the recipient operate in blissful ignorance about the function, utilization and real cost of individual water schemes during implementation. This neglect is hard to explain considering the large efforts spent on collecting information when medium- and long-term plans were prepared (Ch. 10.3). The introduction of a management- and user-oriented information system on donor-assisted projects would be an important first improvement.[5] No massive data collection effort is advocated. It is important that the

---

[4] The need for clearly-stated objectives does not disappear with the use of the adaptive approach to planning and implementation. However, objectives are subject to change, based on the implementation information collected on a continuous basis.

information need at the three levels in the bureaucracy is identified and that the system is geared to those specific needs.Simple is optimal.

A higher degree of *integration between planning and implementation functions* is also needed. Planning and implementation should not be regarded as distinct and separate activities. The isolation of planners from implementors was especially pronounced during the preparation of medium- and long-term plans, as the case studies show. It was a major reason for their non-use. Planning and implementation functions should instead be better linked organizationally – and should not be sharply sequenced over time. Moreover, the emphasis on planning during implementation should be increased in relation to the resources used on pre-implementation planning. In this way information from the implementation process can be fed into the planning process, as already discussed in Ch. 10.3. This may necessitate an organizational change in the RWE and DWE offices (where the two functions are separated as they are at national level, see Figure 2a). It will generally require a more decentralized form of planning. It will certainly also require that donor-assisted planning and implementation activities do not bypass recipient organizations to the extent that has happened in the past. Finally, it would require the involvement of field level staff in certain aspects of operational planning, for which they often have valuable knowledge. Such bureaucratic reorientations are difficult to introduce and can only be done slowly. The reintroduction of District Councils offers a new opportunity for decentralized planning and implementation – and a clear need for central policy direction as argued in Chapter 12.1.

User *participation* in planning and implementation is a third important element in the adaptive approach. The case studies (Chapters 4, 5, 6 and 7) illustrate how difficult it is to transform intentions and plans about participation into reality. Chapter 8 indicates that it is possible to introduce a (limited) degree of participation in rural water sector activities. The unclear Tanzanian policies on the role of beneficiaries in rural water sector policies do, however, need clarification and political approval before these limits can be moved (see Ch. 12.1).

From the perspective of an adaptive approach, three issues are

---

[5] By "management-oriented" is simply meant a system that allows regular comparisons to be made between what was planned and what was actually implemented on issues that are important for decision-making. By "user oriented" is meant a system that allows regular reporting on function (all schemes) and utilization (sample schemes). Beneficiaries would play a major role in providing such data. Ideally the information system should provide a two-way flow between staff and beneficiaries.

important with respect to participation – assuming that the authorities are willing to accept and encourage active participation by the beneficiaries (also when their priorities conflict with the official ones). First, it is necessary to develop planning and implementation procedures that make it possible for the donor and recipient organizations to be responsive to local needs and benefit from local knowledge, and that are conditional on local resource commitment. Procedures are also needed to guide field work and to coordinate the activities of various agencies involved. Second, organizational changes are needed. The ability of beneficiaries to deal with the authorities on water supply issues should be increased by strengthening the capacity of district and village level committees (see below). The coordination of government agencies may require organizational changes too. Finally, staff training in and motivation for a participatory approach must be improved in both recipient organizations and among the technical assistance staff working in this sector.[6] But training is not only needed with respect to participation. It is a key requirement at all levels if the introduction of the adaptive planning and implementation approach is to succeed.

12.3. Increasing Capacity to Plan and Implement from Village Level and Upwards

As explained in Chapter 11, one of the fundamental aims from a donor point of view of employing the adaptive approach to planning and implementation is to increase the capacity of recipient institutions to carry out activities on a sustained basis. This requires that resource allocations to rural activities are adjusted to the capacity of recipient institutions to plan and implement. The pace is therefore likely to be slow. The considerable degree of donor bypassing of local institutions

---

[6] A participatory approach should be context-specific. There are large socio-economic cultural and bureaucratic variations across Tanzania. A specific standard approach is therefore not possible. Furthermore there are many alternative ways in which participator tasks can be located in the recipient organizations. For example, an extension unit function can be carried out in three ways: (i) by the Community Development Staff in the PMO – which will require a coordination of field activities between CD and MAJI; (ii) by Community Development Staff attached to MAJI or specially trained and employed by MAJI; and (iii) by training existing technical staff in MAJI in participatory approaches so that they become multi-purpose workers. Too little experience with different possibilities exists and this prevents specific countrywide recommendations to be made.

which has been practised in the rural water sector has many harmful effects (Ch. 10.5), and the long-term sustainability of donor assistance is not assured in this way, as several case studies show (especially Ch. 4, 5 and 7). It is necessary to change the role of the technical assistance staff towards training with an organizational focus. And to adjust the donor resource inputs toward the capacity of recipient institutions to plan and implement.

Training with an organizational focus is needed at all levels and it should be a key focus of donor assistance. At village level training should aim to increase village capacity to participate in the planning, construction, operation, maintenance and monitoring of schemes. The largest challenge is here to involve women in rural water sector activities. At the district/regional level training should aim to enhance present skills, to develop new ones and to advocate their use in existing or new procedures on the above activities. The same types of skills are needed at the national level. But in addition there is a need to increase the capacity of the Ministry to conduct sector analyses, work out and implement new policies, and coordinate donor assistance. Donors have increasingly withdrawn their direct support to the central Ministry in rural water sector over the last ten years. There is an urgent need to change this. The present lack of central capacity for materials procurement and distribution, and for policy making, coordination and monitoring, is a serious constraint to a decentralized adaptive planning and implementation approach. To ignore central level institutions and to focus on the lower levels only, is a short-sighted donor bias.

It is equally important to adjust donor resource inputs to the rural water sector towards the capacity of recipient institutions to plan and implement. The adaptive planning and implementation approach requires that the knowledge and skills required to plan are gained simultaneously with the capacity to implement plans. Recipient capacity is therefore a key factor in determining the magnitude and type of donor assistance. In the context of rural development assistance government-to-government agreements should be based on field level performance. A slow and gradual implementation rate can therefore be expected.

This leads to a final crucial point. In Chapter 10.5 it was pointed out that the donor preference for the control-oriented approach and the bypassing that it entails, tend to lead to passivity and subterfuge in the recipient organization. Just as villagers have become passive receivers of government-provided services, the recipient organizations have tended to react in the same way to donor assistance in the rural water sector.

A deliberate use of the adaptive planning and implementation approach would hopefully change this. Donor assistance should not significantly and instantly exceed local capacity. Past performance – not only the recipient's targets for the future or the donors' eagerness to move money - should influence the size, trend and composition of donor assistance. This would imply that future assistance is made conditional on past and present use of aid to water sector activities. Not only would this provide an incentive for recipient organizations to perform, it would also transfer the responsibility for project and programme activities to these organizations and the intended beneficiaries. If the benefits of aid are to be sustained, these responsibilities belong nowhere else.

12.4. The Dilemma

No doubt the appeal of the control-oriented approach is caused by its clear division of planning and implementation activities into stages and its strong emphasis on detailed specifications of future activities. If the plans are clear, surely the implementation must comply and the development activities can be controlled? But this is not so, as the case studies show.

In contrast to the control-oriented approach, the adaptive one appears to be an argument for murky generalities. This is a misconception. Compared to the control-oriented approach, the adaptive approach emphasizes:

- the formulation of long-term policies and strategies rather than long-term targets;
- continuous planning linked to implementation, rather than extensive and detailed pre-implementation planning followed by implementations with limited monitoring;
- the regular monitoring and formative evaluation to detect and learn from errors on a continuous basis, rather than periodic external evaluations;
- continuous dialogue with intended beneficiaries to adjust activities to their needs, knowledge and resource commitments, rather than provision of services.

However, in both approaches there is a need for fairly specific short-term objectives to provide direction for implementation. So the adaptive approach to planning and implementation does not imply that detailed planning is not needed or that only small projects will do. The important point is that as projects or programmes mature and grow in size, it may both be possible and necessary to carry out increasingly detailed planning and programming of activities (see Chapter 11.1–2). The crucial thing is that the *initial* period calls for an adaptive approach to planning and implementation, starting with small-scale activities which are then gradually expanded depending on what works and what does not.

A good deal of optimism is required to initiate the changes proposed above and to carry them through. For the results are strongly dependent on the political, economic and social context in which they are introduced. Perhaps the present crisis in Tanzania and the obvious problems of much of the past donor assistance to the rural water sector provide a good occasion to try.

The dilemma that must be confronted in such attempts is this: a faster immediate improvement of the rural water supply situation may result if donors continue to use the control-oriented approach to planning and implementation. But the village level improvements resulting from this approach are not sustainable in the long run. Much slower improvements might result at village level if the adaptive approach were used by donors. But these improvements may perhaps be more sustainable in the long run because they are *also* the result of local commitments and capacity to plan and implement. The challenge for Tanzania is to find the trade-off between buckets full of aid money and buckets full of water!

References

Published

Anderson, I. (1982), "Wells and Handpumps in Shinyanga Region, Tanzania", *Research Paper No. 77*, BRALUP.
Argawala, R. (1983), "Planning in Developing Countries. Lessons of Experience". *World Bank Staff Working Papers*. no. 576.
Arnstein, S.R. (1971), "Eight Rungs on the Ladder of Citizen Participation", in Cohn and Passett (1971).
Balaile, W. (1983), "Rural Water Development and Sanitation in Tanzania: The Rationale of Developing Regional Water Master Plans", in Lium and Skofteland (1983).
Barkan, J.D. and J.J. Okumo (eds.) (1979), *Politics and Public Policy in Kenya and Tanzania*. (New York: Praeger Publishers).
Baum, W. (1982), *The Project Cycle*. (Washington: The World Bank).
Belshaw, D.G.R., (1979), "Regional Planning in Tanzania. The Choice of Methodology for Iringa Region". *Development Division Paper No. 24*; University of East Anglia.
Berry, L. and D. Conyers (1971), "Planning for Rural Development in Tanzania", in BRALUP (1971).
Biswas, A.K. (1981), "Water for the Third World", *Foreign Affairs*, vol. 60, no. 1, pp. 148–166.
Blair, H.W. (1985), "Reorienting Development Administration", *The Journal of Development Studies*, Vol. 21, No. 3, pp. 449–457.
Boesen, J., K. Havnevik, J. Koponen and R. Odgaard (eds.) (1986), *Tanzania. Crisis and Struggle for Survival*. (Uppsala: Scandinavian Institute of African Studies).
Boesen, J., B. Storgård-Madsen and T. Moody (1977), *Ujamaa – Socialism from Above*. (Uppsala: Scandinavian Institute of African Studies).
BRALUP (1971), "Water Supply". Proceedings of the Conference on Rural Water Supply in East Africa, 5–8 April, 1971. University of Dar es Salaam, *Research Paper 20*.
Bryant, C. and L.G. White (1982), *Managing Development in the Third World*. (Boulder, Colorado: Westview Press).
Caiden, N.G. and A. Wildavsky, (1974), *Planning and Budgeting in Poor Countries*. (London: John Wiley & Sons).

Cairncross, S., I. Carruthers, D. Curtis, R. Feachem, D. Bradley and G. Baldwin, (1980), *Evaluation for Village Water Supply Planning.* (Chichester: John Wiley & Sons).
Chagula, W. K. (1971), "Opening Address", in BRALUP (1971).
Chambers, R. (1973), "Planning for Rural Areas in East Africa: Experience and Prescriptions", in Leonard (1973).
Chambers, R. (1983), *Rural Development: Putting the Last First.* (London: Longman).
Clapham, C. (1985), *Third World Politics.* (London: Croom Helms).
Cliffe, L. (ed.) (1967), *One Party Democracy.* (Nairobi: East African Publishing House).
Cliffe, L. and J. Saul, (1973), *Socialism in Tanzania. Policies.* Vol. 2. (Dar es Salaam: East African Publishing House).
Cohn, E.S. and B.A. Passett (eds.) (1971), *Citizen Participation: Effecting Community Change.* (New York: Praeger Publishers).
Cohen, J.M. and N. T. Uphoff (1980), "Participation's Place in Rural Development: Seeking Clarity through Specificity." *World Development,* Vol. 8, pp. 213–235.
Cohen, J.M., M. S. Grindle and S. T. Walker, (1985), "Foreign Aid and Conditions Precedent: Political and Administrative Dimensions". *World Development,* Vol. 13, no. 12.
Crozier, M. (1967), *The Bureaucratic Phenomenon.* (Chicago: The University of Chicago Press).
Dryden, S. (1968), *Local Administration in Tanzania.* (Nairobi: East African Publishing House).
Ekeh, P. (1975), "Colonialism and the two Publics in Africa: a Theoretical Statement". *Comparative Studies in Society History,* Vol. 17, No. 1, pp. 91–112.
Faber, M. and D. Seers (eds.) (1972), *Crisis in Planning.* Vol 1: The Issues, Vol 2: The Experience. (London: Chatto & Windus).
Faludi, A. (1973a), *Planning Theory.* (Oxford: Pergamon Press).
Faludi, A. (1973b), *A Reader in Planning Theory.* (Oxford: Pergamon Press).
Feachem, R. (1980), "Community Participation in Appropriate Water Supply and Sanitation Technologies: The Mythology of the Decade". *Proceedings of the Royal Society.* B series, Vol. 209/1174, pp. 15–29.
Finucane, J.R. (1974), *Rural Development and Bureaucracy in Tanzania. The case of Mwanza Region.* (Uppsala: Scandinavian Institute of African Studies).
Forss, K. (1985), *Planning and Evaluation in Aid Organizations.* (Stockholm: The Stockholm School of Economics).
Fortmann, L. (1980), "Peasants, Officials and Participation in Rural Tanzania. Experience with Villagization and Decentralization". *Special series on Rural Local Organization.* RLO No. 1, Cornell University.
Friedmann, J. (1967), "A Conceptual Model for the Analysis of Planning Behaviour", in Faludi (1973b).

Gow, D.D. and E.R. Morss (1985), "Ineffective Information Systems"; in Morss and Gow (1985).

Gow, D.D. and J. Van Sant (1983), "Beyond the Rhetoric of Rural Development Participation: How Can It Be Done?" *World Development*, Vol. 11, no. 5, pp. 427–466.

Gow, D.D. and J. Van Sant (1985), "Decentralization and Participation: Concepts in Need of Implementation Strategies", in Morss and Gow (1985).

Gray, C. and A. Martens (1983), "The Political Economy of the 'Recurrent Cost Problem' in West African Sahel". *World Development*. Vol. 11, No. 2, pp. 101–117.

Grindle, M.S. (ed.) (1980), *Politics and Policy Implementation in the Third World*. (Princeton, N.J.: Princeton University Press).

Grover, B. (1983), "Water Supply and Sanitation Project Preparation Handbook". Vol. 1, Guidelines. *World Bank Technical Paper Number 12*.

Heaver, R. (1982), "Bureaucratic Politics and Incentives in the Management of Rural Development". *World Bank Staff Working Papers*, No. 537.

Helland-Hansen, E. (1983), "Short Presentation of the Terms of References and Plan Documents for the Water Master Plans Prepared by Consultants from the Nordic Countries for Different Regions in Tanzania", in Lium and Skofteland (1983).

Holmquist, F. (1979), "Class Structure, Peasant Participation and Rural Self-Help", in Barkan and Okumo (1979).

Honadle, G. and R. Klauss (1979), *International Development Administration. Implementation Analysis for Development Projects*. (New York: Praeger).

Honadle, G.H. and J.K. Rosengard (1983a), "Putting 'Projectized' Development in Perspective". *Public Administration and Development*. Vol. 3, pp. 299–305.

Honadle, G., D. Gow, and J. Silverman (1983b), "Technical Assistance Alternatives for Rural Development: Beyond the Bypass Model." *Canadian Journal of Development Studies*. Vol. IV, No. 2, pp. 221–240.

Honadle, G.H., S.T. Walker, and J.M. Silverman (1985), "Dealing with Institutional and Organizational Realities", in Morss and Gow (1985).

Hyden, G. (1973), *Efficiency versus Distribution in East African Cooperatives*. (Nairobi: East African Literature Bureau).

Hyden, G., (1979), "We must run while others walk: Policy Making for Socialist Development", in Kim et al. (1979).

Hyden, G. (1983), *No shortcuts to Progress: African Development Management in Perspective*. (London: Heinemann).

Iliffe, J. (1979), *A Modern History of Tanganyika*. (Cambridge: Cambridge University Press).

International Development Research Centre (1985), *Women's Issues in Water and Sanitation*. Proceedings Series. IDRC-236e. (Ottawa: IDRC).

Johnston, B. and W. Clark (1982), *Redesigning Rural Development: A Strategic Perspective*. (Baltimore: The Johns Hopkins Press).

Killick, T. and J.K. Kinyaua (1980), "On Implementing Development Plans: A Case Study". *ODI-Review*, No. 1.

Kim, K.S., Mabele, R.B. and M.J. Schulteis (eds.) (1979), *Papers on the Political Economy of Tanzania*. (Nairobi: Heinemann Educational Books).

Kleemeier, L. (1984), "Domestic Policies Versus Poverty Oriented Foreign Assistance in Tanzania". *The Journal of Development Studies*. Vol. 20, No. 2, pp. 171–201.

Korten, D.C. (1980), "Community Organization and Rural Development. A Learning Process Approach". *Public Administration Review*. Vol. 40, No. 5, pp. 480–511.

Korten, D.C. and F.B. Alfonso (1983), *Bureaucracy and the Poor. Closing the Gap*. (Singapore: McGraw Hill).

Lele, U. (1975), *The Design of Rural Development*. (Baltimore: The John Hopkins University Press).

Leonard, D.K., (ed.) (1973), *Rural Administration in Kenya*. (Nairobi: East African Literature Bureau).

Lethem, F. and L. Cooper (1983), "Managing Project Related Technical Assistance. The Lessons of Success". *World Bank Staff Working Papers*, No. 586.

Lindblom, C.E. (1959), "The Science of 'Muddling Through'", in Faludi (1973b).

Lindblom, C.E. (1979), "Still Muddling, Not Yet Through". *Public Administration Review*. Vol. 39, No. 6, p. 517–526.

Lium, T. and E. Skofteland (eds.) (1983), *Water Master Planning in Developing Countries*. (Oslo: Norwegian National Committee for Hydrology).

Maeda, J.H.J. (1983), "Creating National Structures for People Enforced Agrarian Development", in Korten and Alfonso (1983).

Mascarenhas, A. (1983), "Determinants for the Implementation and Administration of Water Projects in Tanzania". *Vierteljahresberichte*, No. 94, pp. 335–350.

Miller, D. (1979), *Self-Help and Popular Participation in Rural Water Systems*. Development Centre Studies (Paris: OECD).

Morgan, E.P. (1983), "The Project Orthodoxy in Development: Re-evaluating the Cutting Edge." *Public Administration and Development*. Vol. 3, pp. 329–339.

Moris, J. (1978), "The Transferability of the Western Management Tradition into the Public Service Sectors: An East African Perspective", in *Management Education in Africa: Prospects and Appraisals* (Arusha: East African Management Institute).

Moris, J. (1981), *Managing Induced Rural Development*. (Bloomington: International Development Institute).

Morss, E.R. (1984), "Institutional Destruction Resulting from Donor and Project Proliferation in Sub-Saharan Countries". *World Development.* Vol. 12, No. 4, pp. 465–470.

Morss, E.R. and D.D. Gow (eds.) (1985), *Implementing Rural Development Projects.* (Boulder: Westview Press).

Mujwahuzi, M. (1978), "A Survey of Rural Water Supply Dodoma". BRALUP, Dar es Salaam.

Mukandala, R. (1983), "Trends in Civil Service Size and Income in Tanzania, 1967–1982". *Canadian Journal of African Studies.* Vol. 17, no. 2, pp. 253–264.

Mushi, S.S. (1978), "Popular Participation: The Politics of Decentralized Administration". *Tanzania Notes and Records.* No. 83, pp. 63–100.

Mushi, S.S. (1982), "Aid and Development. An Overview of Tanzania's Experience", in Mushi and Kjekshus (1982).

Mushi, S.S. and H. Kjekshus (eds.) (1982), *Aid and Development. Some Tanzanian Experiences.* (Oslo: Norwegian Institute of International Affairs).

Oakley, P. and D. Marsden (1984), *Approaches to Participation in Rural Development.* (Geneva: International Labour Office).

Paul, S. (1982), *Managing Development Programmes: Lessons from Success.* (Boulder, Colorado: Westview Press).

Paul, S. (1983), *Strategic Management of Development Programmes: Guidelines for Action.* (Geneva: International Labour Organization).

Pfeffer, J. (1981), *Power in Organizations.* (Massachusetts: Pitman Publishing).

Pyle, D.F. (1980), "From Pilot Project to Operational Program in India: The Problems of Transition", in Grindle (1980).

Quick, S.A. (1980), "The Paradox of Popularity: 'Ideological' Program Implementation in Zambia", in Grindle (1980).

Rondinelli, D.A. (1976), "International Assistance Policy and Development Project Administration: The Impact of Imperious Rationality". *International Organization.* Vol. 30, Fall 1976.

Rondinelli, D.A. (1982), "The Dilemma of Development Administration: Complexity and Uncertaincy in Control Oriented Bureaucracies". *World Politics.* Vol. XXV, No. 1.

Rondinelli, D.A. (1983), *Development Projects as Policy Experiments: An Adaptive Approach to Development Administration.* (London: Methuen).

Rondinelli, D.A. (1985), "Development Administration and American Foreign Assistance Policy: An Assessment of Theory and Practice in Aid". *Canadian Journal of Development Studies.* Vol. VI, No. 2, pp. 211-240.

Rweyemamu, A.H.R. and B.U. Mwansasu (1974). *Planning in Tanzania. Background to Decentralization.* (Dar es Salaam: East African Literature Bureau).

Rweyemamu, J.F., J. Loxley, J. Wicken, and C. Nyirabu (eds.) (1972), *Towards Socialist Planning.* (Dar es Salaam: Tanzania Publishing House).

Samoff, J. (1974), *Tanzania. Local Politics and the Structure of Power*. (Madison: The University of Wisconsin Press).
Samoff, J. (1983), "Bureaucrats, Politicians and Power in Tanzania: The Institutional Context of Class Struggle". *Journal of African Studies*. Vol. 10, no. 2, pp. 84–96.
Saul, J., (1972), "Planning for Socialism in Tanzania: The Socio-Political Context"; in Rweyemamu *et al.* (1972).
Saunders, R.J. and J.J. Warford (1976), *Village Water Supply. Economics and Policy in the Developing World*. (Baltimore: The John Hopkins University Press).
Schønborg, R. (1983), "Water Master Planning Coordination Unit (WMPCU). Objectives, Strategies and Perspectives", in Lium and Skofteland (1983).
Shaner, W.W. (1979), *Project Planning for Developing Countries*. (New York: Praeger Publishers).
SIDA – Rapport (1979), "Biståndet och skandalerna". No. 8.
Smith, W.E., F.J. Lethem and B.A. Thoolen (1980), "The Design of Organizations for Rural Development Projects – A Progress Report". *World Bank Staff Working Paper;* no. 375.
Stein, H. (1985), "Theories of the State in Tanzania: A Critical Assessment". *The Journal of Modern African Studies*. Vol. 23, no. 1, pp. 105–123.
Stewart, F. (1986), *Economic Policies and Agricultural Performance. The Case of Tanzania*. (Paris: OECD).
Strachan, H.W. (1978), "Side Effects of Planning and the Aid Control System". *World Development*. Vol. 6, p. 467–478.
Sussman, G.E. (1980), "The Pilot Project and the Choice of an Implementing Strategy: Community Development in India", in Grindle (1980).
Sweet, C.F. and P.F. Weisel (1979), "Process versus Blueprint Models for Designing Rural Development Projects", in Honadle *et al.* (1979).
Therkildsen, O. (1986), "State, Donors and Villages in Rural Water Management", in Boesen *et al.* (1986).
Tordorff, W. (1984), *Government and Politics in Africa*. (London: MacMillan)
Tschannerl, G. (1971), "Introduction", in BRALUP (1971).
United Republic of Tanzania (1968), *Background to the Budget*. (Dar es Salaam. Government Printers).
United Republic of Tanzania (1980), "Mpango wa Maendeleo, 1980–1981". (Dar es Salaam: Government Printers).
United Republic of Tanzania (1981), "Report of the Controller and Auditor-General for the Financial Year ended 30th June, 1980". (Dar es Salaam: Government Printers).
Waide, E.B. (1974), "Planning and Annual Planning as an Administrative Process", in Rweyemamu and Mwansasu (1974).
Warner, D. (1970), "The Economics of Rural Water Supply in Tanzania". Economic Research Bureau, University of Dar es Salaam; *E.R.B. paper*. No. 70.19.

Warwick, D. (1979), "Integrating Planning and Implementation: A Transactional Approach". *Development Discussion Paper*, No. 63, Harvard Institute for International Development, June 1979.

Waterston, A. (1965), *Development Planning. Lessons of Experiences*. (Baltimore: The Johns Hopkins Press.).

Webber, M.M. (1983), "The Myth of Rationality: Development Planning Reconsidered". *Environment and Planning; Planning and Design*. B. Vol. 10, p. 89–99.

White, G.F., D.J. Bradley and A. U. White (1972), *Drawers of Water. Domestic Water Use in East Africa*. (London: The University of Chicago Press).

Widstrand, C. (1980), "Water Conflicts and Research Priorities". Water Development, Supply and Management, Vol. 8; *Water and Society*, Conflicts in Development, Part 2.

Wijk-Sijbesma, C. (1985), "Participation and Women in Water Supply and Sanitation: Roles and Realities". *Technical Paper no. 22* (The Hague: International Reference Centre).

Wingaard, B. (1983), "Opening Address", in Lium and Skofteland (1983).

World Health Organization (1985), "Catalogue of External Support", Publication nr. 7, December 1985; *International Drinking Water Supply and Sanitation Decade*.

World Health Organization (1986b), "The International Drinking Water Supply and Sanitation Decade. Review of Regional and Global Data (As at 31st December 1983)". *WHO Offset Publication*. No. 92.

Unpublished

AIB (1979), "Action Programme in the Rural Water Supply Programme in Tanzania". Report to the Ministry of Water, Energy and Minerals, December, 1979.

Anon (1984), "Study Team Report on System Design and Implementation of the Rural Water Supply Programme for Morogoro and Shinyanga Region". Morogoro.

Arnzen, B. (1984), "Report on Organizational and Managerial Issues". Report on HESAWA, September 1984.

Athanari, M., B. Sedin, P.E. Thomsen, and B. Wingård (1983), "Review of the Water Master Planning Coordination Unit at Maji, Ministry of Water and Energy, Tanzania", March 1983, Dar es Salaam.

Ausi, H.M. (1979), "Rural Water Supplies and Regional Development. A Case Study of the Shinyanga Wells Programme in Bariadi District, Shinyanga Region, Tanzania". Master's Thesis (The Hague: Institute of Social Studies).

Boesen, J. (1986), "Aiming Too High, Planning Too Much and Achieving Too Little". Centre for Development Research, Copenhagen, A 86.5.

Bonnier, C.J. (1980), "Shallow Wells in Tanzania; Introduction and Discussion Points"; in Ministry of Water, Energy and Minerals (1980).

Book, A. (1984), "A study of Operation and Maintenance of the Water Supply Programme in the Regions Kagera, Mwanza and Mara in Tanzania". Report by short-term consultant.

BRALUP/CDR (1982), "Socio-Economic Studies", *Water Master Plans for Iringa, Ruvuma and Mbeya Regions*, Vol. 12. Bureau of Resource Assessment and Land Use Planning, University of Dar es Salaam. Centre for Development Research, Copenhagen.

Brokonsult (1978a), *Water Master Plan for the Mara, Mwanza and West Lake Regions. Summary*. Draft final report.

Brokonsult (1978b), *Water Master Plan for the Mara, Mwanza and West Lake Regions. Vol. 3*. Draft final report.

Brokonsult (1978c), *Water Master Plan for the Mara, Mwanza and West Lake Regions. Vol. 12*. Draft final report.

Brokonsult (undated). Water Master Plan for Tabora Region. Vol. 3.1. Engineering.

CCKK (1981), "Workshop on Integrated Planning". Summary of discussions; Mbeya, 20–21 July, 1981.

CCKK (1982a), "Introduction"; *Mbeya Water Master Plan*, vol. 1.

CCKK (1982b), "Summary"; *Mbeya Water Master Plan*, vol. 3.

CCKK (1982c), "Village Water Supply Studies"; *Mbeya Water Master Plan*, vol. 4A.

CCKK (1982d), "Village Water Supply. Implementation"; *Mbeya Water Master Plan*, vol. 4B.

DANIDA (1978), "Report from Joint Tanzanian/Danish Water Appraisal Mission". October 1978.
DANIDA Steering Unit (1982 onwards), "Progress Reports", Half-yearly and Yearly reports. DANIDA Steering Unit, Ubungo, Dar es Salaam.
DANIDA (1984a), "Vandprojektet i Tanzania", DANIDA informerer.
DANIDA (1984b), "Implementation of Water Master Plans for Iringa, Mbeya and Ruvuma Regions". Report prepared by the Annual Joint Tanzanian/Danish Review Mission, February 1984.
DANIDA (1985b), "Implementation of Water Master Plans for Iringa, Mbeya and Ruvuma Regions". Report prepared by the Annual Joint Tanzanian/Danish Review Mission, October 1985.
DANIDA (1985), "Project Guidelines: Appraisal; Planning". March 1985.
DANIDA (1986), "Implementation of Water Master Plans for Iringa, Mbeya and Ruvuma Regions". Report by a Review Mission on Economy, Organization and Administration, April 18, 1986, Dar es Salaam.
DHV (1979), "Morogoro Wells Construction Project. First Progress Report". January 1979. DHV Consulting Engineers.
DHV (1980), "Organizational Aspects of a National Shallow Wells Programme in Tanzania". December 1980, National Shallow Wells Programme.
DHV (1981), "Morogoro Wells Construction Project. Sixth Progress Report". August 1981, DHV Consulting Engineers.
DHV (1982), "Morogoro Wells Construction Project. Ninth Progress Report". December 1982, DHV Consulting Engineers.
DHV (1983), *Water Supply Survey, Southern Morogoro Region*. October 1983, DHV Consulting Engineers.
DHV (1984), "Final Report. Rehabilitation". Morogoro Wells Construction Project, September 1984, DHV Consulting Engineers.
Engstrom, J.E. and J.E. Wann (1975), "Inventory of Rural Water Supply Projects in Tanzania". December 1975, Dar es Salaam.
FINNIDA (1984), "Tanzania: Mtwara-Lindi Rural Water Supply Project". Report of the Evaluation Mission, March 1984.
Finnwater (1977a), "Mtwara-Lindi Water Master Plan. Main Report". March 1977.
Finnwater (1977b), "Mtwara-Lindi Water Master Plan. Annex L. Water Development Programme. Lindi Region". March 1977.
Finnwater (1977c), "Mtwara-Lindi Water Master Plan. Annex I. Water Supplies Mtwara Region." March 1977.
Finnwater (1980), "Rural Water Construction Project in Mtwara and Lindi Regions." Phase I. Final Report, January 1978 – March 1980.
Finnwater (1982), "Mtwara-Lindi Rural Water Project." Phase II, Final Report, April 1980 – December 1981.
Finnwater (1982 onwards), "Mtwara-Lindi Rural Water Project." Phase III. Various Quarterly Progress Reports, January 1982 – October 1984.

Finnwater (1985a), "Project Document of Phase IV". Undated.

Finnwater (1985b), "Mtwara-Lindi Water Master Plan. Revision. Part: Water Supply". Interim Report, April 1985.

Hannan Andersson, C. (1985), "Domestic Water Supply Improvements in Tanzania. Impact on Rural Women". Dar er Salaam.

HESAWA (1984), "Hesawa. Programme Progress Report". Nov. 1984.

Hordijk, A., E.M. Munuo, D. Ricardo and M. Schröcle (1982), "Evaluation of the Netherlands sponsored Water Projects in Morogoro Region – Tanzania". May 1982, Ministry of Foreign Affairs: DGIS; Holland.

Hyden, G., P.H. Mallya, N. Mtalo and H.J. Nyundo (1973), "Expatriate Effectiveness in Development Management in Tanzania: A Case Study of the Ministry of Water Development and Power". Report prepared by a special study team for Min. of Water Development and Power, October 15th, 1973.

IRC (1980), "Project Profile for the Development of a Community Participation Component in the Tanzanian Rural Water Supply Programme". IRC/CEP/80.5; November 1980.

IRC (1981), "Project for the Development of a Community Participation Component in the Tanzanian Rural Water Supply Programme. Report of a first Mission to the United Republic of Tanzania". IRC/CEP/TA.01; April 1981.

IRA/CDR (1983), "Socio-Economic Studies. Village Participation on Water and Health", *Water Master Plans for Iringa, Ruvuma and Mbeya Regions*, Vol.13. Institute of Resource Assessment, Dar es Salaam; Centre for Development Research, Copenhagen.

IRA/WMPCU (1984), "HESAWA. Principles and Procedures for Community Participation, Health Education and Sanitation". Report, April 1984.

Jennings, A. (1981), "Aid Modalities in the United Republic of Tanzania". Study prepared for Review Meeting in Addis Abeba, 4–15 May, 1981.

Kauzeni, A.S. and J.H. Konter (1981), "Institutionalization of Shallow Wells under Tanzanian Administration". BRALUP.

Kleemeier, L. (1982), "Decentralized Planning in Tanzania. A History and Evaluation of the Role of Foreign Assistance 1971–82". Paper presented at the Political Science Seminar, University of Dar es Salaam, November 29, 1982.

Mattoke, W.T. (1984), "Summary of Various Party Policies/ Statements on Rural Development in Tanzania". Unpublished Report, February 1984, Kivikoni College, Dar es Salaam.

Ministry of Water, Power and Electricity (1972), "Hotuba ya Waziri Kuhusu Makadirio ya Mwaka 1972/73". Statement of the Minister during presentation in the National Assembly.

Ministry of Water, Energy and Minerals (1977), "Summary of the Meeting of Regional Water Engineers, Regional Geology and Mines Officers with Officials of the Ministry of Water, Energy and Minerals"; held in Dodoma from 13th April 1977 to 16th April 1977. May 1977.

Ministry of Water, Energy and Minerals (1980), "Morogoro Conference on Wells", Proceedings; Mikumi Wildlife Lodge, 18–22 August 1980. Report, November, 1980.

Ministry of Water, Energy and Minerals (1980a), "Summary Report on the Annual Conference for Regional Water Engineers and Ministry Officials", Tanga, May 6 – May 9, 1980.

Ministry of Water and Energy (1983), "Community Participation and Water Master Planning in Tanzania. Background and Present Situation". WMPCU No. 31. July 1983.

Ministry of Water and Energy (1983), "Estimates of Expenditures for the Year 1983/84". Statement of the Minister during Presentation in the National Assembly.

Ministry of Water and Energy (1984), "Shallow Wells Programme. Final Report". Shallow Wells Technical Committee, Dar es Salaam, February, 1984.

Ministry of Water and Energy and Australian Development Assistance Bureau (1984), *Evaluation of the Tanzania Village Water Development Project.* Final Report, April 1984.

Ministry of Water and Energy/FINNIDA (1981), "Mtwara-Lindi Regional Water Development Programme. Evaluation Report". May 1981.

Msimbira, N.K. (1984), "Rural Water Supply Programme". Paper presented to the Regional Water Engineers' Conference, 27–31 August, 1984. Ministry of Water, Energy and Minerals.

Mwapachu, H.B. (1980), "Opening Speech", in Ministry of Water, Energy and Minerals (1980).

NEDECO (1974), "Sociology." *Shinyanza Water Supply Survey. Water Master Plan Study for Shinyanza Region.* Final Report, Vol. D.

Nordberg, A. (1985), "Recommendations for the HESAWA Programme in Tanzania." Report, April 1985.

PEU (1984a), "Preliminary Evaluation Report on the Water Component of the Mwanza Rural Development Project". April 1984, Report No. PEU 3/84. Planning and Evaluation Unit, Prime Minister's Office, Mwanza.

PEU (1984b), "Report on Survey of Shallow Wells installed by Water Component of the Mwanza Rural Development Project". April 1984, Report No. PEU/84. Planning and Evaluation Unit, Prime Minister's Office, Mwanza.

PMO/IRC (1984), "Project for the Development of a Community Participation Component in the Tanzanian Rural Water Supply Programme". Final Report, July 1984. Prime Minister's Office; International Reference Centre.

Rigsrevisionen (1983), "Beretning til Statsrevisorerne om Danmarks Projektbistand til Tanzania" (RB 1102/83). København; December 1983.
Ringelberg, J. (1980), "Maji Shallow Well Programme, Mwanza"; in Ministry of Water, Energy and Minerals (1980).
Rimer, O. (1970), "Tanzania Rural Water Supply Development". Ministry of Agriculture, Food and Cooperatives; Water Development and Irrigation Division. April 1970.
Samset and Stokkeland Consulting (1984), "HESAWA, Lake Regions, Monitoring and Evaluation." July 1984.
Schluter, M. (1982), "An Analysis of Budgetary Allocations". September 1982, World Bank, Washington.
SIDA (1983a), "Implementation plans. Lake Regions. Terms of Reference." Project document.
SIDA (1983b), "Insatspromemoria. Fortsatt Stöd till Tanzanias Landsbygdsvattenprogram." 24. February, 1983.
SIDA (1984), "Water Strategy: Water Supply Programmes for Rural Areas." Policy Document.
SIDA/PMO (1983), *Rural Water Sector Review in Tanzania*. Report from the joint Swedish/Tanzanian Team, April 1983.
SIDA/PMO (1984a), *Rural Water Sector Review in Tanzania*. Report from the joint Swedish/Tanzanian Team, February 1984.
SIDA/PMO (1984b), *HESAWA Programme Review*. Report from the joint Swedish/Tanzanian Team, November 1984.
TANU (1971), "Mwongozo", Party Guidelines. National Executive Committee, Dar es Salaam.
TISCO (1980), "Assessment on Rural Water Schemes in Dodoma and Lake Regions, Tanzania." TISCO, Dar es Salaam.
VIAK (1981), "Implementation Plan. Rural Water Supply. Mwanza." Final report, August 1981.
Van der Laak, F.H.J. (1980), "Organization and Maintenance", in Ministry of Water, Energy and Minerals (1980).
Warner, D. (1973), *Evaluation of the Development Impact of Rural Water Supply Projects in East African Villages*. Report EEP-50, Stanford University.
WHO/World Bank, (1977), "Rural Water Supply Sector Study. Tanzania." WHO/World Bank Cooperational Programme.
World Bank (1977a), "The Sukuma: a Socio-Cultural Profile." Working Paper C-1, Tanzania: Mwanza/Shinyanga Rural Development Project.
World Bank (1977b), "Water Supply Component." Working Paper C-7, Tanzania: Mwanza/Shinyanga Rural Development Project.
World Bank (1977c), "Village Self-Help Program." Tanzania; Mwanza/Shinyanga Rural Development Project.
World Health Organization (1986a), "International Drinking Water Supply and Sanitation Decade. Mid-Decade Progress Review." Report by the Director General. WHO, 21st March, 1986; paper A39/11.

List of Abbreviations

AFYA	Ministry of Health
AIB	Allmänna Ingenjörsbyrån, Sweden
BRALUP	Bureau of Resource Assessment and Land Use Planning, University of Dar es Salaam (later IRA)
CCKK	Carl Bro; Cowiconsult, Kampsax; Krüger; Copenhagen, Denmark
CCM	Chama Cha Mapinduzi
CD	Community Development Department, PMO
CDR	Centre for Development Research, Copenhagen
DANIDA	Danish International Development Agency
DDD	District Development Director
DGIS	Directorate-General for International Cooperation, Ministry of Foreign Affairs, Holland
DHV	Dwars, Heederik en Verhey; Amersfoort, Holland
D.Kr.	Danish Kroner
DWE	District Water Engineer
FINNIDA	Finnish International Development Agency
GoT	Government of Tanzania
HESAWA	Health, Sanitation, Water Project funded by SIDA
IDRC	International Development Research Centre, Canada
ILO	International Labour Office
IMF	International Monetary Fund
IO	Implementation Office, Iringa, Mbeya and Ruvuma
IRC	International Reference Centre for Community Water Supply and Sanitation, Holland
IRA	Institute of Resource Assessment, University of Dar es Salaam (previously BRALUP)
MAENDELEO	Community Development Department, PMO
MAJI	see Ministry
MDWSP	Morogoro Domestic Water Supply Plan
MINISTRY	Denotes the ministry responsible for rural water sector development
MRDP	Mwanza Rural Development Project
NSWP	National Shallow Wells Programme
O & M	Operation and Maintenance
OECD	Organization for Co-operation and Development
PARTY	see CCM and TANU

PEU	Project Evaluation Unit, Mwanza
PMO	Prime Minister's Office
PMU	Project Management Unit
RDD	Regional Development Director
RSCM	Regional Steering Committee Meeting
RWE	Regional Water Engineer
RWMP	Regional Water Master Plan
SEC	Socio Economic Study
SIDA	Swedish International Development Authority
SU	Steering Unit, DANIDA, Dar es Salaam
TANU	Tanganyika African National Union
TISCO	Tanzania Industrial Studies and Consulting Organization
TOR	Terms of Reference
Tsh	Tanzania Shillings
UK	United Kingdom
UN	United Nations
UNDP	United Nations Development Programme
UNICEF	United Nations Children's Fund
URT	United Republic of Tanzania
VPC	Village Participation Coordinator
VSHP	Village Self-Help Project Unit, Mwanza
WB	World Bank
WHO	World Health Organization
WMPCU	Water Master Planning Coordination Unit